Understanding Structures

Analysis, Materials, Design

Derek Seward
Department of Engineering
University of Lancaster

Third Edition

palgrave

D1394373

First edition 1994
Second edition 1998
Third edition 2003
PALGRAVE MACMILLAN
Houndmills, Basingstoke, Hampshire RG21 6XS and
175 Fifth Avenue, New York, N.Y. 10010
Companies and representatives throughout the world

PALGRAVE MACMILLAN is the global academic imprint of the Palgrave Macmillan division of St. Martin's Press, LLC and of Palgrave Macmillan Ltd. Macmillan® is a registered trademark in the United States, United Kingdom and other countries. Palgrave is a registered trademark in the European Union and other countries.

ISBN–13: 978–0–333–97386–8
ISBN–10: 0–333–97386–0

This book is printed on paper suitable for recycling and made from fully managed and sustained forest sources.

A catalogue record for this book is available from the British Library.

10	9	8	7	6	5	
12	11	10	09	08	07	06

Printed in China

Acknowledgements

The author and publishers wish to thank the following for permission to use copyright material.

BSI Standards for extracts. Complete editions of the standards can be obtained by post from BSI Customer Services, 389 Chiswick High Road, London W4 4AL.

The Appendix on standard rolled steel sections contains data from *Steelwork Design: Guide to BS 5950 1: Volume 1: Section Properties and Member Capacities* and is reproduced by kind permission of the Director, The Steel Construction Institute.

Cover design photographs of the Humber Bridge and the Forth Road Bridge were kindly provided by Acer Consultants Limited.

Contents

Preface

This book explains the fundamentals of structural analysis, materials and design. These topics are often treated as separate subjects, but it is my belief that it is better to introduce all three topics in an integrated fashion. This enables the student to tackle realistic design problems in the shortest possible time, and, because he or she can see the relevance of the theory, it produces good motivation. It is also a way of providing early exposure to the actual **process** of design – from initial concept through to final details. Although much of the book deals with the design of individual structural elements in various materials, it also considers the way these elements are used in complete structures, which is the essence of structural form. To attempt to cover such a wide range of material in a modestly sized book has not been easy. In selecting material for inclusion I have given preference to those issues which occur most frequently in the real world of structural design.

The book is intended for students of civil and structural engineering, building, architecture and surveying. It is of use at both first year degree and BTEC level. Because it contains much real design data, it may also be useful as a work of reference for the non-specialist practitioner.

In order to produce safe and economic structures, a large part of the structural design process is inevitably numerical. However, where possible, the book avoids a mathematical approach. The aim is to develop a 'feel' and awareness for the physical behaviour of structures. As the title of the book suggests, the emphasis is placed on understanding. For this reason a large number of illustrations are used to portray structural behaviour. A minimum of mathematical knowledge is required – largely simple algebra and trigonometry. Where, as in certain proofs of formulae, slightly more advanced mathematics is unavoidable, it is presented in such a way that uninterested readers can avoid it without affecting their wider understanding of the book. The scope of the book is restricted to statically determinate structures, in the belief that a thorough understanding of these is required before moving on to structures which contain redundant members.

The treatment is in line with the latest limit state approach, now adopted by national standards for structural design in most materials. However, mention is also made of the more traditional permissible stress approach, which is still in common use for some materials. An unusual feature of the book is the inclusion of a chapter on structural loads. This vital stage in the design process is often difficult for the beginner, and yet is rarely covered in text books. A more controversial feature is the way that the book explains ultimate plastic bending strength before the elastic case. The basic design method adopted in national standards for the principal structural materials now assumes that full plastic strength can be developed. Indeed, run-of-the-mill design of steel and reinforced concrete beams can now take place without any knowledge of elastic theory. Plastic strength is also easier to understand. It is only for historical reasons that elasticity is usually taught before plasticity.

Each stage of the design process is illustrated by a realistic numerical example which is based on genuine design data. It is hoped that after reading this book the student will have developed a real skill for structural design, as well as sharing in the satisfaction, pleasure and excitement of this highly creative process.

Derek Seward

Symbols

A	cross-sectional area
A_c	area of concrete
A_e	effective area
A_s	area of tensile reinforcement
A_{sc}	area of compressive reinforcement
A_{sv}	area of two legs of shear link
a	pitch angle of roof, length of plate in a cross-section
B	beam width
b	width of concrete
C	width of column
D	beam depth, depth of arch
d	effective depth of reinforcement
d_s	depth of slab
E	modulus of elasticity
e	eccentricity of load
F	axial force
F_C	compressive force
F_{CC}	compressive force in concrete
F_{CS}	compressive force in steel
F_T	tensile force
f_{cu}	concrete characteristic strength
f_k	characteristic compressive strength of masonry
f_{kx}	characteristic tensile bending strength of masonry
f_y	reinforcement characteristic strength
f_{yv}	characteristic strength of shear links
G	shear modulus
G_k	characteristic dead load
H	horizontal component of force
I	second moment of area
j	number of joints
k_a	active earth pressure coefficient
k_p	passive earth pressure coefficient
L	length, length of span
L_E	effective length
M	bending moment
M_F	moment due to prestressing force
m	number of members

m_1	moment at node 1
N	number of stress cycles, ultimate load capacity of a reinforced concrete column
P	concentrated point load, column load,
P_{crit}	critical buckling load
P_{1x}	force at node 1 in x direction
p	hydrostatic pressure
p_b	bending strength
p_c	compressive strength
Q	first moment of area about neutral axis
Q_k	characteristic imposed load
q	ground pressure under foundation
q	shear flow
R	radius of curvature, resultant force, radius of shaft
R_{AV}	vertical reaction force at A
r	radius of gyration
S	plastic section modulus, magnitude of stress fluctuation
s_v	spacing of shear links
T	torque
t	beam web thickness, wall thickness
V	vertical component of force, shear force
v	design shear stress, Poisson's ratio
v_c	concrete shear strength
W_k	characteristic wind load
w	magnitude of uniformly distributed load
x	an arbitrary or unknown distance, neutral axis depth in concrete, torsional index
Y	distance from beam surface to neutral axis
y	distance from neutral axis
Z	elastic section modulus
z	depth of liquid or soil, lever-arm in concrete beam
γ	unit weight, shear strain
γ_f	partial safety factor for loads
γ_m	partial safety factor for material strength
Δ	deflection of structure
δ	extension of member
δ_{1x}	displacement of node 1 in x direction
ε	strain
ϕ	angle of twist
θ	angle
θ_1	rotation at node 1
μ	slip factor
σ	stress
σ_y	yield stress
σ_u	ultimate stress
τ	shear stress

Chapter 1

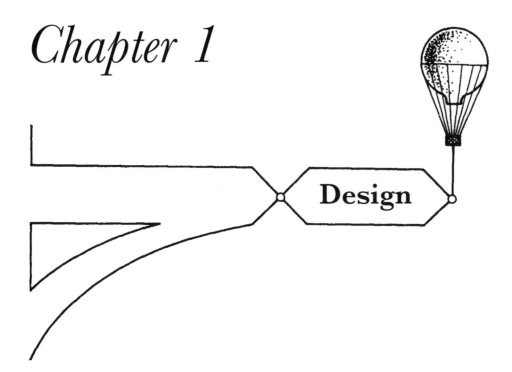

Design

Topics covered

Definition of a structure
Difference between analysis and design
The design process
National standards
Permissible stress design
Limit state design

1

1.1 Introduction

This book is about the **design** and **analysis of structures** in the commonly used structural materials. This chapter starts by looking at what we mean by these words, and then goes on to explain the process that designers perform in order to successfully evolve solutions to structural problems. Finally two distinct approaches or 'design philosophies' are explained – the traditional **permissible stress** method, and the more modem **limit state** method.

An important point to make at the outset is that structural design is a highly creative activity, and not just a case of plugging numbers into formulae. It is also an activity which must be done responsibly, in order to produce **safe** structures. In mechanical engineering most manufactured products with a structural component, like cars for example, can be repeatedly tested to destruction to prove the design at the prototype stage. Modifications can then be made before the product is put on the market. With most civil engineering and building structures there is no prototype. **The design must be right first time**. These structures can also have a very long working life of a hundred years or more, and it is the responsibility of the designer to ensure that it remains safe throughout its projected life.

Figure 1.1
A building structure safely transmits loads down to earth

1.2 Structure

The word **structure** can be used to describe any organised system, such as the 'management structure of a company' or the 'structure of the atom'. However, for the purposes of this book we will limit the definition to the following:

'A structure is a system for transferring loads from one place to another.'

In the case of **building structures** this often means transferring the load of people, furniture, the wind etc. (as well as the self-weight of the building itself) safely down to the foundations and hence into the ground (*figure 1.1*).

Also in this book we are limited to the consideration of **manmade** structures. Nature can show us many superbly efficient examples of structures which have evolved to support loads. Some of these are shown in *figures 1.2a–c*, and although natural

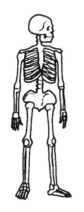

Figure 1.2a
The human skeleton is a structure which maintains the shape of the body, keeps the various organs and muscles in the right place and transmits loads down to the ground

structures can be analysed and investigated using the methods described in this book, we are really interested in the exciting task of creating new structures.

1.3 Analysis

The word **analysis** is generally taken to mean the process whereby a **particular** structure with **known** loads is investigated to determine the distribution of forces throughout the various members that make up that structure. Also it includes determining the distribution of stresses within individual members, which result from the forces imposed on them. Finally, it covers the calculation of deflections (i.e. how far the structure will move) under a particular set of loads. The analysis of a structure is necessary to prove that it is strong enough to support a given set of loads.

Analysis tends to be based on mathematics, and the aim is to get as close as possible to the uniquely correct solution. Analysis is a vital **part** of the design of safe and cost-effective structures, however, it cannot take place until the basic form of the structure has been decided. We firstly need to settle such questions as 'should we use steel or concrete?', 'How many supports to the beam should we provide?' These early decisions are referred to as **design** and not analysis.

1.4 Design

Design is a more difficult word to define than analysis because it means such different things to different people. The term **designer** is commonly used to describe people who design patterns on carpets, or determine the shape of motor car bodies. Whilst aesthetic design is very important it is **not** the subject of this book.

Engineering design or **structural design**, which **is** the subject of this book, is equally creative – just consider the brilliance of the Forth Railway Bridge or the Golden Gate Suspension Bridge (*figure 1.3*), two very different solutions to the similar problem of building a long-span bridge over water.

Even within our own field the word 'design' can be used in two very different ways. Firstly it is used to describe the whole creative process of finding a safe and efficient solution to an engineering problem. Consider, for example, the above case of designing a

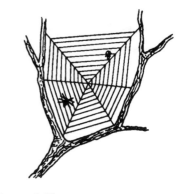

Figure 1.2b
The spider's web is a good example of a **tension** structure. The weight of the spider and its prey is supported by the tensile strength of the web

Figure 1.2c
Shells are particularly efficient structures – being rigid and lightweight. The curved surfaces can be very thin

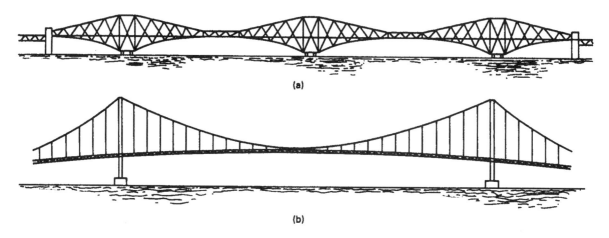

(a)

(b)

Figure 1.3
Two brilliant, but very different, solutions to a similar problem

(a) The Forth Railway Bridge, Edinburgh, Scotland, 1890

(b) The Golden Gate Bridge, San Francisco, USA, 1937

Figure 1.4a
Swing fixed to a tree

Figure 1.4b
Tubular steel swing

bridge across a river. There is of course no 'correct' answer, but out of the infinite number of possible solutions, some are clearly much better than others. It is the job of the designer to come up with the best solution within available resources. The design process is dealt with in more detail in the next section.

The second way that we use the word 'design' is in a more restricted sense. It refers to the activities, which often come after the analysis stage, when the forces in each structural member are known. It is the process of determining the actual size of a particular steel column or the number of reinforcing bars in a reinforced concrete beam. This is referred to as *element design*. Determining the number of bolts in a connection or the length of a weld is called *detail design*. There may be several choices open to the designer, but element design and detail design are not as creative as the broader design process. They are, however, very important, and bad detail design is a major cause of structural failures.

1.5 The design process

The best way to explain the general approach adopted by a designer in solving an engineering problem is to take a simple example – in this case the design of a child's garden swing. Clearly, with a small project like this, the formal procedure outlined here would not be consciously followed. The designer would simply use his or her judgement to determine the shape of the swing support structure and the size of the members. Nevertheless it illustrates the steps necessary for the successful completion of larger projects.

The usual starting point for a project is a **client** who has a **requirement**, and the client will approach the designer, who is often a consulting engineer or architect, and commission him or her to produce a design which satisfies the requirement. The first thing that the designer will request from the client is a clear description of exactly what is wanted. This is called the **design brief**.

In our case the client is a child who requires a garden swing, and the brief is:

'Provide, within a period of two weeks, a garden swing, which will be a play facility for many years.'

To arrive at a satisfactory solution the designer will go through the following stages:

- **Stage 1** is the **site survey and investigation**. Suitable locations will be examined and checked to ensure that there are no obstructions within the arc of the swing. Problems of access must be considered and the ground conditions in wet weather taken into account. Finally a few trial-holes might be dug to check that any rock is not too close to the surface to prevent foundations being excavated.

- **Stage 2** involves investigating **alternative structural concepts**. There are an enormous range of structural shapes and materials which could be used and a few of these are shown in *figures 1.4a–f*. Each of these must now be evaluated against clear rational criteria in order to select just one or two schemes for further detailed consideration. The type of factors that would be considered in relation to each scheme are as follows:

(a) Fixed to tree. This is the cheapest and best solution, but there is no convenient tree.

(b) A tubular steel frame. They are available from the supermarket. Little labour is involved and the swing is mobile. There are some concerns about cost, durability and appearance, but it is worth bearing in mind.

(c) Suspended from a helium balloon. An interesting solution, but totally impractical.

(d) Welded and bolted steel beams. Very robust but it would have to be prefabricated by a local welder. The result is exceedingly ugly and would require painting to prevent rust.

Figure 1.4c
Suspended from a helium balloon

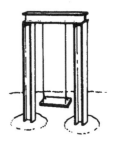

Figure 1.4d
Heavy steel sections

Figure 1.4e
A timber-framed structure

Figure 1.4f
Swing fixed to a wall

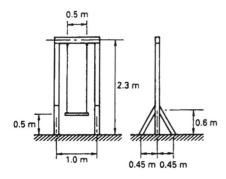

Figure 1.5
Basic dimensions

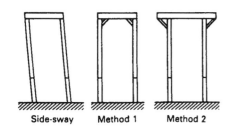

Figure 1.6
The problem of side-sway and two possible methods of preventing it

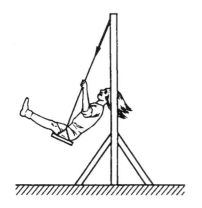

Figure 1.7
Inclined load acting at the top of the swing frame

(e) Timber frame with concrete foundations. Good appearance and relatively cheap but high labour content and some maintenance required for good durability.

(f) Timber and steel bracket fixed to wall. Unless it is possible to fix the swing away from the corner of the wall there will be safety problems.

The choice therefore comes down to either (b) from the supermarket or (e) the timber frame. We will assume that after due investigation of available products the decision is made to go for the timber frame.

- **Stage 3** involves more **detailed development** of the selected scheme. The basic overall dimensions must be decided (*figure 1.5*). A method of bracing the frame to prevent side-sway must be found (*figure 1.6*). The type of timber to be used will probably be decided at this stage and the range of available sizes investigated.

- **Stage 4** is the **assessment of loads** on the structure. This is not such a straightforward task as you might at first think. The following issues need to be settled.

1. Should the swing be designed solely for the weight of a child, or should it be assumed that at some time during its life an adult may use it? I would suggest the latter.

2. We need to increase the load by a percentage to allow for the fact that the load is **dynamic**, i.e. the load is moving, and we must take account of the effects of its inertia. Also the load may be suddenly applied if someone jumps onto the swing. Increasing the static load by, say, 100% ought to be enough to compensate for these effects.

3. The load is not always applied vertically downwards. When the load is inclined, as shown in *figure 1.7*, it causes much more severe bending and overturning problems for the structure. Wind loads often cause severe lateral loads, although they are not much of a problem with thin skeletal structures such as this.

4. Additional forces result from the actual self-weight of the structural members. With structures such as large bridges these forces are usually much bigger than those that result from the traffic passing over the bridge. These self-weight loads present a problem at this stage of the design process because we do not yet

know the final size (and hence weight) of the structural members. The designer must therefore make intelligent estimates which should be checked later. Obviously, with increased experience the designer gets better at making these estimates. In our case the effect of self-weight will be very small and could be ignored.

- **Stage 5** is the **analysis** of the structure. Firstly we must transform the diagram of the proposed structure to obtain a simplified structural model which is amenable to analysis. This chiefly involves classifying the joints between the members as either pinned or continuous. This topic is dealt with in more detail later in the book. Our structural model is shown in *figure 1.8*.

The loads are now applied to the structural model and the frame analysed to determine the forces and the amount of bending in each member. It may be necessary to carry out more than one analysis for different load cases. For the swing structure we would probably do it twice – firstly with the load vertical and secondly with the load inclined to produce the worst overturning effect.

- **Stage 6** is **element design** and **detail design**. Each member must be considered in turn. Taking into account the forces obtained from the analysis, the required size of the member is calculated so that acceptable stresses are not exceeded. Careful thought must be given to the connections so that forces are adequately transmitted from one member to another. The designer must check all the possible ways in which the structure could fail. Some of these are shown in *figure 1.9* and you may be able to think of some others.

Finally, with real structures, the designer must clearly communicate his requirements to the builder by means of detailed drawings and specifications.

All stages of design are summarised in *figure 1.10*. It is not always possible to separate the process into such neat steps and often analysis and element design will proceed in parallel. Also the analysis to calculate the displacement of the structure can only take place after the member sizes are known.

1.6 National standards

Because the safety of structures is so important, the major industrial countries produce guidance documents, **codes of practice**,

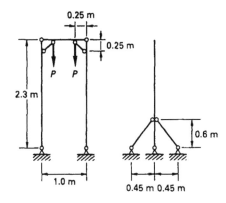

Figure 1.8
The structural model

Figure 1.9a
Seat breaks

Figure 1.9b
Rope snaps

Figure 1.9c
Bracing connections fail

Figure 1.9d
Beam breaks

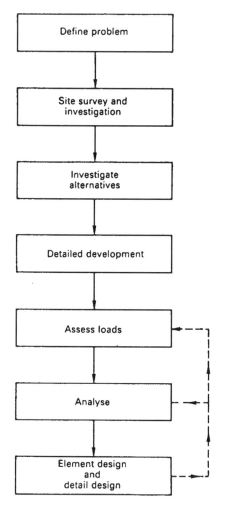

Figure 1.10
The stages of design

containing the wisdom and experience of eminent researchers and practising professionals. There tends to be a separate document for each structural material. In the UK these are British Standards or BSs, in the USA they are ASTM standards and in Germany they are DIN standards. In Europe the process of replacing the separate national codes by single **Eurocodes** is well advanced. Each Eurocode is, however, produced with a series of National Application Documents (NADs). These contain factors and constants to ensure that the code is tuned to the national practice of each country.

These documents contain rules of a very general nature, and they should always be complied with where they are relevant. However, because they are so general, they do require interpretation by qualified designers who understand the underlying theory. They do not replace sound experience, understanding and judgement.

Very small structures, such as those involved in individual houses, are covered by national or local Building Regulations. These contain simple rules covering such issues as the minimum thickness of walls for certain heights. If designers stay within the stated limits then no calculations are required. However, designers are at liberty to go beyond the limits provided that they can justify it by calculation and reference to the relevant standards.

It is not intended that this book should explain particular codes of practice in any detail; however, the theories and procedures presented generally comply with the major national and international standards.

1.7 Design philosophies

Clearly the loads which will act on a structure throughout its life are clearly very difficult to predict at the design stage, and we must accept that even the best of estimates may be in error. Likewise the strength of many structural materials is difficult to predict – particularly those that involve naturally variable materials such as timber, masonry and concrete. It follows that sensible design procedures must allow for these uncertainties by including factors of safety. A simple definition of a factor of safety is:

$$\text{factor of safety} = \frac{\text{load to cause failure}}{\text{actual load on structure}}$$

Clearly the value of the factor of safety must be greater than 1 to prevent collapse. The magnitude of the factor of safety that we should aim for depends upon the predictability of both the loading and the material properties, but can range from about 1.5 for steel structures up to 5.5 for brickwork.

The main difference between the two approaches described in the following two sections is the way in which adequate factors of safety are achieved.

1.7.1 *Permissible stress design*

Permissible stress design (sometimes called allowable stress design) is the method formerly advocated by the national standards for all materials. It has now largely been replaced by the following limit state approach. It is included here because it is still in widespread use by new designers and it remains the accepted approach for mechanical engineering components.

In essence, the permissible stress codes of practice provide the designer with a maximum stress value that can be used for a particular material. This permissible stress is **not** the failure stress for the material, but is reduced to provide a factor of safety. As an example, designers are instructed to limit the stress in a mild steel beam to a permissible value of 165 N/mm^2, whereas we know that it can actually withstand about 275 N/mm^2 before yielding. The designer must ensure that the permissible stress is not exceeded at any profit in a structure. This process is intended to provide an adequate factor of safety to compensate for errors in **both** loading and material strength. The designer may, however, be unaware of the precise value of the factor of safety.

Apart from collapse, there are other defects which can render a structure inadequate, such as floor beams which sag too much under load, or cracks in concrete which are too wide. These problems must be dealt with in addition to checking the stresses.

The main criticism of this approach is the fact that the designer cannot vary the factor of safety to compensate for the degree of uncertainty concerning the loads.

1.7.2 *Limit state design*

This approach was developed on the continent of Europe and has now been accepted throughout the world. The aim is to be clearer and more scientific about the aims of the design process.

We have seen how a structure can be in such a state that it is unfit for use – e.g. excessive deflections or cracking. The point at which it is unacceptable is known as a **limit state**. The aim of the design process is therefore:

'To ensure that a structure has an acceptable probability of not reaching a limit state throughout its life.'

This seems to be largely stating the obvious, although the use of the word probability is highly significant. It implies that we are aware that in the real world we must accept some failures from time to time. Clearly some limit states are more important than others – the most important obviously being **collapse**, and this is termed the **ultimate limit state**. All the others, such as excessive deflections, are termed **serviceability limit states**. The limit state approach allows us to set a higher probability against reaching the ultimate limit state than the other less important serviceability limit states.

In practice, the principal way in which the limit state approach differs from the permissible stress approach is that the factors of safety are explicitly stated, and applied separately, firstly to the loads and secondly to the ultimate strength of the material. This process is clearly illustrated in the following example.

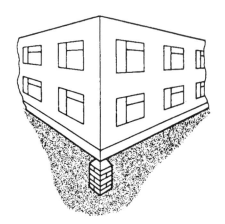

Figure 1.11a
A brick column supporting the corner of a warehouse

Example 1.1

Figure 1.11a shows the corner of a warehouse building which is supported on a short brickwork column. The load on the column is estimated to be as follows:

Load from weight of the building itself (**dead load**)   = 600 kN
Load from contents of the warehouse (**imposed load**) = 450 kN

The column is constructed from **Class B engineering bricks** with mortar made from 3 parts sand to 1 part cement.

Determine the required cross-sectional area of the column if the design is based on:
a. Permissible stress principles to CP 111: 1970 (now withdrawn).
b. Limit state principles to BS 5628: 1978.

Solution

Comment – Do not worry too much about the details at this stage – the purpose of this example is to illustrate the different approach to safety factors.

a. Permissible stress design

$$\text{Total load on column} = 600 + 450 = 1050 \text{ kN}$$

Comment – Class B engineering bricks have a guaranteed minimum crushing strength of 48.5 N/mm² as individual bricks. However, when built into a brickwork wall the mortar joints produce a considerable reduction in crushing strength. CP 111 gives the permissible basic stress for brickwork made with class B engineering bricks and 3 to 1 mortar as:

$$\text{Permissible basic stress} = 3.3 \text{ N/mm}^2$$

Comment – This figure already contains a safety factor to take account of errors in estimating the loads and also possible variations in material strength. As we shall see in the next chapter – stress = force area or area = force stress.

Answer Area required $= \dfrac{1050 \times 10^3}{3.3} =$ **318 200 mm²**

b. Limit state design

Comment – The first stage is to apply a load factor (γ_f) to the loads. Limit state design recognises the fact that it is easier to predict the dead loads than the imposed loads. Therefore a safety factor of 1.4 is applied to the dead load and a higher factor of 1.6 to the imposed load. The factored load is called the **design load**.

$$\begin{aligned} \text{Design dead load} &= 1.4 \times 600 = \quad 840 \text{ kN} \\ \text{Design imposed load} &= 1.6 \times 450 = \underline{\quad 720 \text{ kN}} \\ \text{Total design load} &= 1560 \text{ kN} \end{aligned}$$

Comment – The strength given in a limit state code for a particular material is known as the **characteristic strength**. *It does not contain a safety factor. We must therefore apply a safety factor, γ_m, to the material strength.*

BS 5628 is more complicated than most standards in recommending a factor which varies between 2.5 and 3.5 depending upon the degree of quality control used. We will assume that the highest standards of quality control are used for both the manufacture of the bricks and the site construction work – hence γ_m = 2.5.

Characteristic compressive strength = 15 N/mm^2

(from BS 5628)

$$\text{Design strength} = \frac{15}{2.5} = 6.0 \text{ N/mm}^2$$

Answer: Area required $= \dfrac{1560 \times 10^3}{6.0} =$ **260 000 mm^2**

Comments:

1. The two methods have produced roughly the same result with the permissible stress method requiring an area of column 22% greater. If we had not assumed special quality control in the limit state method and consequently increased the material factor, γ_m, to 3.5, then the limit state area would be the bigger.

2. It is clear that the permissible stress approach is simpler and quicker than the limit state approach, but it does not provide as much scope for logically fitting the design to the prevailing conditions. Some designers argue that the permissible stress approach should have been retained for simple jobs.

*3. In practice the actual size of the column would be **rounded-up** to suit the modular dimensions of bricks. The arrangement of bricks shown in figure 1.11b has a cross-sectional area of 305 300 mm^2. It would therefore be adequate for limit state design, but not for permissible stress design.*

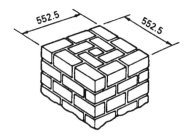

552.5   552.5

Figure 1.11b
A suitable brick arrangement for limit state design

1.7.3 *Accuracy of structural calculations*

It is becoming clear that the design of structures requires many assumptions to be made about the nature of the loads acting on a structure, the strength of the structural materials used and the form of the simplified structural model. All of these factors mean that there is absolutely no point in presenting calculations with too much precision. Even the simplest scientific calculators give a minimum of eight figures, and it is the mark of an amateur to copy them all down in structural calculations. Three, or at most four, significant figures are generally adequate.

On many occasions a designer needs to round off a calculated value of a dimension to suit a commonly available size. For example calculations may indicate that a steel plate needs to be 9.7 mm thick. In this case it would be specified as 10 mm, which is a standard size that can be obtained 'off the shelf'. A difficult decision is sometimes needed if the calculations indicate a thickness of say 10.2 mm. Should a thickness of 10 mm be specified or the next size up of 12 mm? In general it is prudent to play safe and round up. In this way, if the calculations are in accordance with a particular standard, it can be said that the design fully complies with that standard.

1.8 Summary of key points from chapter 1

1. For the purposes of this book a **structure** can be defined as **a system for transferring loads from one place to another**.
2. **Design** is a widely used term that is applied to everything from say the creative conceptual design of a bridge to the detail design of a bolted connection. 'Analysis' is the mathematical part of the design process that is concerned with the determination of forces, stresses and deflections within a given structure.
3. In general all structural design should comply with national or international standards and should only be carried out by competent and qualified designers.
4. **Permissible** or **allowable stress design** tries to ensure that at no time during the working life of a structure will the stresses exceed a safe value which is always less than the actual strength of the material. This approach has been largely superseded.
5. Current practice uses **limit state design** which is based on probabilities. To guard against reaching the **ultimate limit**

state of **collapse**, individual safety factors are applied to both the loads (i.e. the loads are multiplied by γ_f) and the material strength (i.e. the material strength is divided by γ_m).

1.9 Exercises

E1.1 *Section 1.2* showed examples of natural structures. Bearing in mind that the purpose of a structure is to transfer loads from one place to another, draw simple sketches of three structures either taken from other examples in nature and/or from examples in the home such as furniture.

E1.2 *Figure 1.3* showed two very different design concepts for a large-span bridge. Sketch two different concepts for either:

a) A tall tower to support a television transmission mast, or

b) the roof of an Olympic sports stadium.

E1.3 A farmer requires an access bridge over a stream (about 5.5 metres span and 2.5 metres roadway width). The bridge will be used for farm vehicles and animals. Describe the six stages of design as illustrated in section 1.5 for a garden swing. Use plenty of simple sketches to illustrate your answer.

a) In **stage 1** describe the factors that would be taken into account in siting the bridge. Produce a simple cross-section, drawn roughly to scale, showing the banks of the stream and the water level.

b) Suggest and discuss the advantages and disadvantages of at least three alternative concepts in **stage 2** before selecting one for further development.

c) In **stage 3** produce a roughly dimensioned side elevation and cross-section of your chosen concept.

d) In **stage 4** try to think of some of the loads that the bridge must be designed to carry.

e) For **stage 5** transform your sketch of the bridge into a simple line diagram with either pinned or continuous joints. Ignore further analysis.

f) Finally, for **stage 6**, show at least three ways in which your bridge could fail.

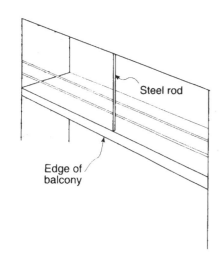

Figure 1.12
Balcony supported by steel rod

E1.4 *Figure 1.12* shows a balcony structure supported by a tensile steel rod. The force in the rod is as follows:

Load from self-weight of balcony (dead load) = 12 kN
Load from people (imposed load)   = 16 kN

Determine the required cross-sectional area of the rod based on each of the following two design philosophies:

a) Permissible stress design:
   permissible stress for steel in tension   = 150 N/mm²

b) Limit state design:
   characteristic ultimate stress for steel   = 275 N/mm²
   partial safety factor for material strength, γ_m = 1.0
   partial safety factor for dead load, γ_f = 1.4
   partial safety factor for imposed load, γ_f = 1.6

a) *permissible stress area = 187 mm²*
b) *limit state area = 154 mm²*

Chapter 2

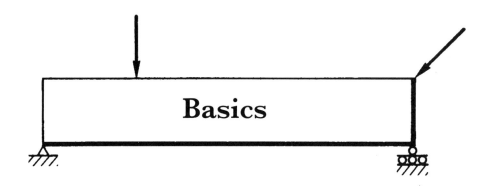

Topics covered

Units
Forces – vectors, components, resultants
Moments
Equilibrium
Reactions and types of support
Stress
Strain

2.1 Introduction

For most readers this chapter will be revision of material previously covered. It can be described as simple **statics**. This means that we are only concerned with the mechanics of forces applied to bodies which are not moving. They are said to be in **statical equilibrium**. Although much of this material may be familiar, it is worth ensuring that you fully understand it because it is fundamental to the rest of the book.

Figure 2.1
Because of gravity, objects with the same mass only exert a sixth of the force on the moon

2.2 Forces

The loads acting on a structure have **mass** which is usually measured in kilograms (kg). For the design of static structures on earth it would be possible to work in kg throughout – but things are never that simple! In order to be consistent with physics and mechanics, we must convert the mass into a **force**.

The basic unit of force is the **newton (N)**. The force exerted on a structure by a static load is dependent upon both its mass and the **force of gravity**. The mass of a body, measured in kg, is a fundamental measure of quantity and is related to the number of atoms of material present. If you consider the beam shown in *figure 2.1*, it would have exactly the same mass on earth as it would have on the moon. The force that it exerts on the man's shoulder, however, would be six times greater on earth than on the moon because of the earth's greater gravitational pull. Gravity is measured by the acceleration (g) that it imparts to a falling body, and on earth can be taken as 9.81 m/s^2.

From **Newtons second law of motion**:

$$\text{force} = \text{mass} \times \text{acceleration}$$

and $\qquad$ 1 newton = force to accelerate 1 kg at 1 m/s^2

Therefore, on earth, the force exerted by a mass of 1 kg is:

$$\text{force} = 1 \times g = 9.81 \text{ N}$$

The force exerted by a body as a result of gravity can be described as its **weight**. However, weight can also be used less formally to mean mass.

Figure 2.2
A newton is roughly the force exerted on earth by an (stationary) apple

To develop some feel for the magnitude of a newton, it is easy to remember that it is roughly the weight of an apple (*figure 2.2*). For many structural loads this is rather a small unit; therefore it is common to use the kilonewton (kN):

$$1 \text{ kN} = 1000 \text{ N}$$

Again, to give some idea of the magnitude of a kilonewton:

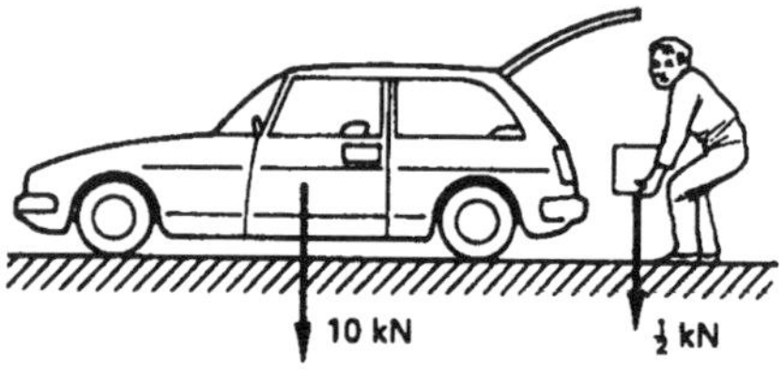

- 0.5 kN is the weight of a full bag of cement, which is about the limit for the average person to lift (*figure 2.3*).

- 10 kN is about the total force exerted on the road by a small car (*figure 2.3*). This has roughly a mass of 1 tonne (1 tonne = 1000 kg).

To summarise, if we wish to convert a mass (kg) to a force (N) we multiply by 9.81. If the result is large we can then divide by 1000 to get kN.

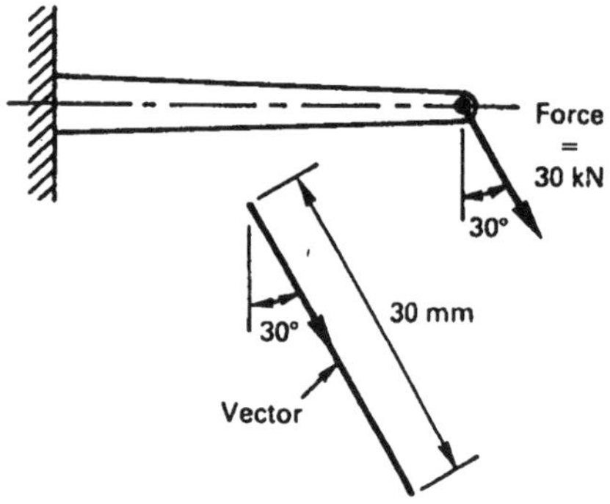

Figure 2.3
Half a kilonewton (0.5 kN) is about the limit for the average person to lift and 10 kN is the force exerted on the road by a small car

2.2.1 *Vectors*

Because a force has both a magnitude and a direction it is a **vector** quantity. It is often convenient to represent force vectors graphically by a line whose **length** is proportional to the **magnitude** of the force and whose **direction** is **parallel** to the force.
Figure 2.4 shows a force acting on a structure and its vector drawn to a scale of 1 kN = 1 mm.

Figure 2.4
A force vector

2.2.2 *Resolving forces*

We can frequently simplify problems by splitting a single force into its **components** in say the vertical and horizontal directions – this is called **resolving**.
Figure 2.5 shows the previous force vector resolved into its vertical component, V, and its horizontal component, H. Either by scale drawing or from simple trigonometry it can be seen that:

$$V = 30 \times \cos 30° = 26.0 \text{ kN}$$

$$H = 30 \times \sin 30° = 15.0 \text{ kN}$$

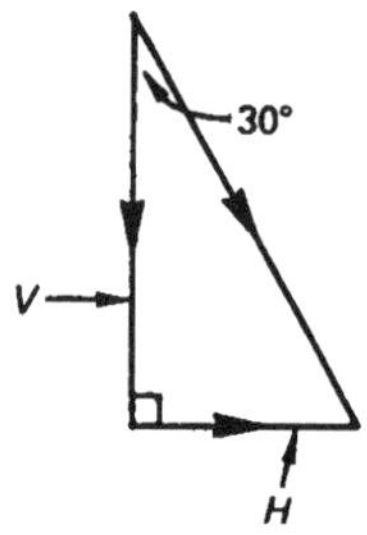

Figure 2.5
Forces can be resolved into vertical and horizontal components

Thus, so far as our original structure is concerned, the 30 kN load could be replaced by its two components as shown in *figure 2.6*. The effect on the structure is identical.

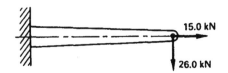

Figure 2.6
A force can be replaced by resolving it into its components without changing the effect on the structure

2.2.3 *Resultant forces*

The previous process can be reversed, and two or more forces can be combined to produce their **resultant**. The vectors of each force are simply added (in any order) and the resultant is the vector required to close the circuit.

This is shown in *figure 2.7a* for two forces, which produce a **force triangle**, and in *figure 2.7b* for several forces, which produce a **force polygon**. Note that the vectors of the original forces must be drawn so that their directions form a continuous string, whereas the direction of the resultant force opposes the flow. (Some readers may be familiar with resolving two forces by drawing a **parallelogram of forces**. The same result is obtained.)

An alternative approach is to use trigonometry instead of drawing. Firstly the forces must each he resolved into two perpendicular directions (usually vertical and horizontal). We then find the net sum of all the vertical components, followed by the net sum of all the horizontal components. Pythagoras's theorem can then be used to find the length of the diagonal, which is the required resultant force. For the two forces shown in *figure 2.7a* this would be as follows:

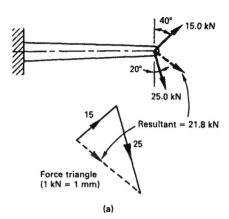

$$\text{vertical components} = 25 \cos 20° - 15 \cos 40°$$
$$= 23.5 - 11.5 = 12 \text{ kN } \textit{downwards}$$

$$\text{horizontal components} = 25 \sin 20° + 15 \sin 40°$$
$$= 8.55 + 9.64 = 18.2 \text{ kN } \textit{to the right}$$

$$\textbf{Resultant} = \sqrt{(12^2 + 18.2^2)} = \textbf{21.8 kN}$$

$$\text{Angle to vertical} = \tan^{-1} \frac{18.2}{12} = 56.6°$$

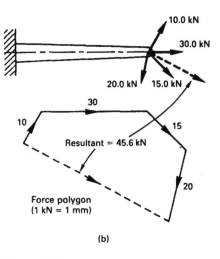

Figure 2.7
Multiple forces acting at a point may be replaced by their resultant

2.3 Moments

If the line of action of a force does not act directly **through** a particular point, then it exerts a **moment about** that point. The magnitude of a moment about a point is the value of the force multiplied by the perpendicular distance from the line of action of the force to the point. In other words the distance is the shortest distance between the line of action and the point under consideration. This is shown in *figure 2.8b*.

Thus   Moment = force × perpendicular distance

Therefore the units of moment are **Nm** or more frequently **kNm**. *Figure 2.8a* shows a force acting perpendicular to the axis of a member.

$$\text{Moment of } P \text{ about the point } X = P \times L$$

To talk about the moment of P, without specifying the point about which moments are taken, is meaningless.

Figure 2.8b shows a force inclined at an angle θ. To find the moment of this force about the point X it is necessary to calculate the perpendicular distance from the line of the force to the point X.

$$\text{Moment of } P \text{ about the point } X = P \times L \cos \theta$$

As an alternative to the above we could firstly resolve the force into its perpendicular and parallel components:

$$\text{Perpendicular component, } V = P \cos \theta$$

$$\text{Parallel component, } H = P \sin \theta$$

From *figure 2.8c*:

$$\text{Moment of } P \text{ about the point } X = P \cos \theta \times L \text{ as before.}$$

Note that because the line of the parallel component, H, passes through the point X, it does not affect the moment. Remember, a force only produces a moment about a point when the line of action of the force acts at a distance from a point. In this case the distance is zero.

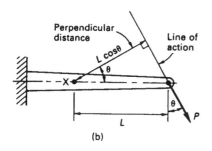

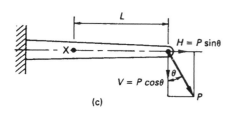

Figure 2.8
The moment of a force about a point is equal to the magnitude of the force multiplied by the perpendicular distance

2.4 Equilibrium and reactions

Any structure subjected to loads must be provided with supports to prevent it from moving. The forces generated on the structure by these supports are called **reactions**. If the structure is in **equilibrium** (i.e. not moving) then the **net** forces from the loads and reactions must be zero in all directions. This simply means that a structure with say total downward loads of 100 kN must be supported by upward reactions of 100 kN. Also there must be no out-of-balance moments about **any** point.

2.4.1 *Reactions from vertical loads*

If a structure is subjected to only vertically downward loading then in general the net support reactions will be vertically upwards. An exception is with arch structures and these are dealt with in *chapter 12*. For rigid bodies two basic independent equilibrium equations can be used:

- summing vertical forces
- taking moments.

The term 'taking moments' exploits the fact that for **any** point on a structure the sum of all the clockwise moments must equal the sum of all the anti-clockwise moments (otherwise the structure would spin round). The point selected for taking moments about is usually at one of the unknown reactions. Remember that a moment is a force times a perpendicular distance. Thus when taking moments about a point **all** distances must be measured from that point.

Because only two independent equations are available we can generally only solve problems which involve two unknown vertical support reactions. Structures which contain more supports must be analysed by more advanced techniques, which are beyond the scope of this book.

Example 2.1

Figure 2.9a shows a bridge deck, of weight 500 kN, supporting a heavy vehicle weighing 300 kN. Suppose we wish to find the value of the support reactions at A and B when the load is in the position shown.

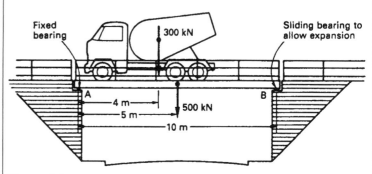

Figure 2.9a
A bridge with a heavy vehicle

Solution

The first stage is to convert the diagram of the bridge into a structural model which can be analysed. This is shown in *figure 2.9b*. Note that the fixed bearing has been replaced with a pinned support which will allow rotation to take place but **no** lateral movement. The sliding bearing has been replaced by a roller which permits both rotation **and** lateral movement. This combination is known as **simply supported**. Types of support are dealt with more fully in *section 2.5*.

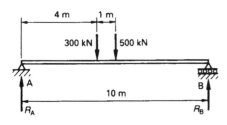

Figure 2.9b
The above structure can be reduced to this simplified structural model for analysis

Analysis

Take moments about A to find the unknown vertical reaction R_B. (From now on this will be written in the form 'ΣM about A to find R_B'. Σ is the Greek capital letter **sigma** and means 'the sum of'.) The sum of the clockwise moments are equated with the sum of the anticlockwise moments.

$$(500 \times 5) + (300 \times 4) = R_B \times 10$$

clockwise   anticlockwise

Answer $R_B = 370$ **kN**

Comment – Make sure you understand where the above figures come from. Each bracket contains a moment which consists of a force times a distance, and all the distances are measured from the point A.

We could now ΣM about B to find R_A, but an easier way is to consider vertical equilibrium – written $(\Sigma V = 0)$. That is the sum of all vertically downward forces must equal the sum of all vertically upward forces.

$$500 + 300 = 370 + R_A$$

downwards   upwards

Answer $R_A = 430$ **kN**

Comments:

1. Representing the reactions by upward forces may seem confusing at first. Remember that we are investigating the equilibrium of **the beam**. *The forces must therefore be those experienced by the beam. If we now wished to consider the forces acting on the bridge abutments* **from the beam** *the directions would be reversed (figure 2.9c).*

2. Clearly if the load had been precisely in the centre of the span we could have concluded from symmetry and vertical equilibrium that $R_A = R_B = (300 + 500)/2 = 400$ kN.

3. In this case the load was represented by a single force located at the centre of gravity of the vehicle. If we had been presented with a series of three individual axle loads, then obviously we would have had extra terms when taking moments, but the result would have been the same.

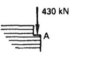

Figure 2.9c
The forces on the bridge abutments are equal but opposite in direction to the reaction forces on the beam

Example 2.2

Figure 2.10a shows a worker, of mass 70 kg, standing on the edge of an elevated access platform. The plank she is standing on weighs 8 kg. Find the plank support reactions.

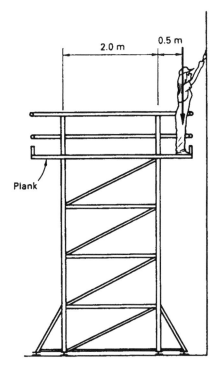

Figure 2.10a
An elevated access platform

Solution

Loads

The first step is to convert the loads into forces and produce the structural model.

$$\text{weight of worker} = 70 \times 9.81 = 687 \text{ N}$$
$$\text{weight of plank} = 8 \times 9.81 = 78 \text{ N}$$

The structural model is shown in *figure 2.10b.*

Comment – Clearly the weight of the plank is evenly spread throughout its length (known as a uniformly distributed load). However, for the purposes of this example we will, quite correctly, consider its weight acting as a point load concentrated at its centre of gravity. Uniformly distributed loads are dealt with in more detail in chapter 4.

Analysis

($\sum M$ about A to find R_B)
$(78 \times 1.0) + (687 \times 2.5) = R_B \times 2.0$

Answer $R_B = 898 \text{ N}$

($\sum V = 0$) $78 + 687 = 898 + R_A$

Answer $R_A = -133 \text{ N}$

Comments:

1. The negative value for R_A implies that the original direction shown in figure 2.10b is wrong and that the plank must be held down to prevent it from tipping.

2. It can be seen in figure 2.10b that the supports to the plank are again shown as a roller and a pin. This is of little significance when all the loads are vertical, but we will see in the next section that it is important with non-vertical loads.

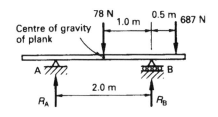

Figure 2.10b
The structural model of the plank

2.4.2 *Reactions from non-vertical loads*

If a structure is loaded by horizontal or inclined loads then we cannot assume that the support reactions are simply vertically upwards. One of the supports must provide a horizontal reaction component in order to satisfy horizontal equilibrium. Another basic equilibrium equation is available to add to the other two:

- summing vertical forces
- taking moments
- summing horizontal forces.

This enables us to evaluate three unknown support reactions including one which is horizontal. This explains why the previous examples have shown structures with one roller and one pin as supports. If a structure had two pinned supports then an extra horizontal reaction is introduced – making four unknowns, i.e. two vertical and two horizontal at each support. Evaluating the reactions then requires a more advanced method of analysis, taking into account the stiffness of the structure, which is beyond the scope of this book.

Example 2.3

Figure 2.11a shows an anchorage for the cable of a ski-lift. The cables produce a total pull of 220 kN at the top of the anchorage. Determine the support reactions.

Solution

Loads

It is usually convenient to resolve inclined loads into their vertical and horizontal components.

$V_L = 220 \sin 30° = 110$ kN
$HL = 220 \cos 30° = 191$ kN

The structural model is as shown in *figure 2.11b*.

Analysis

By definition a roller can only resist forces perpendicular to the plane of its rollers. Any horizontal forces must therefore be resisted by the pinned support at A.

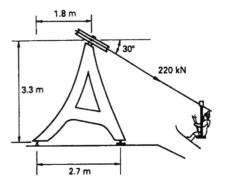

Figure 2.11a
A ski-lift anchorage

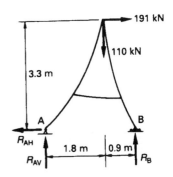

Figure 2.11b
The structural model of the ski-lift anchorage

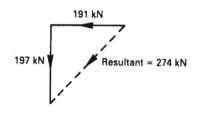

Figure 2.11c

The horizontal and vertical reactions at A could be replaced by the resultant force inclined as shown in the force triangle

$(\Sigma H = 0)$ $R_{AH} = 191$

left   right

Answer $\mathbf{R_{AH} = 191\ kN}$

$(\Sigma M$ about A to find $R_B)$
$(191 \times 3.3) + (110 \times 1.8) = R_B \times 2.7$
Answer $\mathbf{R_B = 307\ kN}$

$(\Sigma V = 0)$ $110 = 307 + R_{AV}$
Answer $\mathbf{R_{AV} = -197\ kN}$

Comment – In the above we have considered the reaction at A to be resolved into its horizontal and vertical components. It can be equally well represented as the resultant, as shown in figure 2.11c.

2.5 Supports

In order to be able to analyse a structure it is necessary to be clear about the forces that can be resisted by each support. *Figure 2.12* shows the principal types of support and the number of reactive forces provided by each

A **fixed** support can resist vertical and horizontal forces as well as a rotational moment. This means that a structure only requires one fixed support to satisfy all three equations of equilibrium. In practice a beam which is rigidly built-in to a massive concrete wall may be considered to have a fixed support. However, the support must be rigid enough to prevent any rotation if it is to be considered truly fixed.

A **pinned** support can resist vertical and horizontal forces but not a moment. This means that a single pinned support is not sufficient to make a structure stable. Another support must be provided at some point to prevent the structure from rotating about the pin. In practice, many structural connections are assumed to be pins, even if a specific point of rotation is not provided. Thus simple bolted connections in a steel frame are often considered to be pinned connections for the purposes of design. In reality a small moment may be resisted, but it is ignored.

A **roller** support can provide only a single reaction force, which is always perpendicular to the axis of the rollers. A good

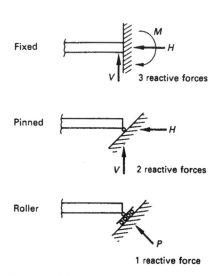

Figure 2.12

Types of support for structures

combination of supports for a structure is therefore a pin plus a roller. Between them they provide enough support to satisfy the three equations of equilibrium. Roller supports must be provided for very long structures such as bridge beams so that large forces are not produced by thermal expansion and contraction. They can take the form of rubber bearings, which are designed to permit lateral movement.

Supports to **real** structures are usually approximated to one of the above.

2.6 Stress

It is important to be able to distinguish between **external forces** – such as applied loads, and **internal forces** which are produced in structural members as a result of applying the loads. Thus if a person weighing say 70 kg were to dangle from a vertical rope, they would be applying an external force to the rope of $70 \times 9.81 = 68.67$ N. However, for vertical equilibrium the rope must contain an internal tensile force of 68.67 N (assuming the rope is weightless). Internal forces cause **stress**.

$$stress = internal\ force\ per\ unit\ area$$

therefore $$\mathbf{stress = \frac{force}{area}}$$

Thus stress can be thought of as the **intensity** of internal force. Stress is usually symbolised by σ, the Greek lower case letter **sigma**. The **units of stress** are commonly taken to be **N/mm²**. Another unit which may be used is the **pascal** (Pa).

$$1\ pascal = 1\ N/m^2$$

This is, however, an inconveniently small quantity and most stresses would be measured in mega-pascals (1 MPa = 10^6 Pa). In this book we will use N/mm² throughout which is consistent with most of the national standards.

An external force, P, is axially applied to a structural member, of cross-sectional area A, as shown in *figure 2.13a*. Make an imaginary cut through the member and consider the equilibrium of the remaining piece shown in *figure 2.13b*. Clearly there must be an internal force, P, to satisfy vertical equilibrium.

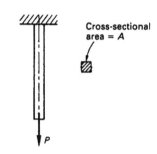

Figure 2.13a
An axial tensile force applied to a member

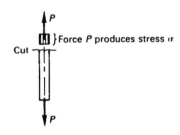

Figure 2.13b
An internal force must balance the external load to satisfy vertical equilibrium

Figure 2.13c
A member subject to compression

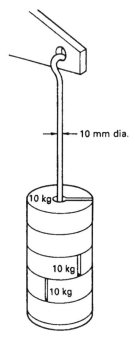

Figure 2.14a
A weight-hanger for applying loads

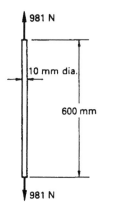

Figure 2.14b
The structural model of the weight-hanger

$$\text{stress in member, } \sigma = \frac{P}{A}$$

In this case the force and the stress are **tensile**.
In *figure 2.13c* the member contains **compressive** force and stress.
Because the forces are applied axially (i.e. on the centroid of the member) the above stresses are **uniform** right across the whole cross-section.

Example 2.4

Consider the structure shown in *figure 2.14a* which is a weight-hanger for applying loads to a piece of laboratory equipment. The steel rod is 600 mm long and has a diameter of 10 mm. Calculate the stress in the rod when the hanger is fully loaded by ten 10 kg weights. Ignore the self-weight of the hanger itself.

Solution

Load

$$\text{Total force on hanger} = 10 \times 10 \times 9.81 = 981 \text{ N}$$

The structural model is shown in *figure 2.14b*.

Analysis

$$\text{Cross-sectional area} = \frac{\pi \times 10^2}{4} = 78.5 \text{ mm}^2$$

$$\text{stress} = \frac{\text{force}}{\text{area}}$$

Answer $$\text{stress, } \sigma = \frac{981}{78.5} \quad \mathbf{12.5 \ N/mm^2}$$

2.7 Strain

In common conversation the words stress and strain are often taken to be synonymous; however, they do have very distinct meanings, and it is vital that you understand the difference. Whereas stress is a measure of the load on each square millimetre of material, **strain is a measure of how much each millimetre length of the material deforms under stress**. It is therefore related to the stiffness of the material rather than the strength. Thus when someone says 'can it take the strain?' they usually mean 'can it take the stress?'.

$$\text{strain, } \varepsilon = \frac{\text{change in length}}{\text{original length}} = \frac{\Delta L}{L}$$

This means that strain is the change in length per unit length. Strain is usually symbolised by ε, the Greek lower case letter **epsilon. The units of strain are dimensionless**. Because strains are usually small they are often presented as a percentage.

The property which determines how much strain occurs for a given stress is the **modulus of elasticity (E)** or **Young's modulus**. This tends to be a constant for a particular material.

$$E = \frac{\textbf{stress}}{\textbf{strain}}$$

As the units of strain are dimensionless, the units of modulus of elasticity are clearly the same as stress i.e. **N/mm^2**.

From above $\qquad \text{strain} = \dfrac{\text{stress}}{E}$

We can see from this that materials with high E values will have relatively small strains and can be described as **stiff**. Some examples of E values are given in the next chapter.

Example 2.5

Consider again the weight-hanger in *example 2.4*. If the steel rod has a modulus of elasticity, E, of 205 000 N/mm^2, determine:

a. the percentage strain in the rod.
b. the extension of the rod.

Solution

From *example 2.4* stress in rod = 12.5 N/mm^2

From above   strain, $\varepsilon = \dfrac{\text{stress}}{E}$

$$= \dfrac{12.5}{205\ 000} = 61.0 \times 10^{-6}$$

Answer   % strain = 6.1 $\times$ 10^{-3}

From   strain, $\varepsilon = \dfrac{\text{change in length}}{\text{original length}}$

extension = strain $\times$ original length

$$= 61.0 \times 10^{-6} \times 600$$

Answer   extension = 0.0366 mm

Comment – The small value of the extension reflects the fact that the rod is fairly short and is only stressed to about 1/20th of the value required to break the rod. Also steel has a high E value compared to many other materials.

2.7 Summary of key points from chapter 2

1. For structural calculations loads given in kilograms must be converted into newtons:
$$1 \text{ kg} = 9.81 \text{ N}$$
2. A force can be represented by a **vector** whose length is proportional to the magnitude of the force and whose direction is parallel to the force.
3. A force, P, inclined at $\theta°$ to the vertical can be **resolved** into vertical and horizontal components:
   vertical component   = P $\times$ cos θ
   horizontal component = P $\times$ sin θ
4. Two or more forces acting at a point can be combined into the **resultant** force by constructing a force polygon.

5. If a force, P, does not act through a point then it exerts a **moment** about that point. The units of a moment are kNm.
   Moment = force × perpendicular distance
6. **Stress**, σ = internal force per unit area   (N/mm^2)
   $$\sigma = P/A$$
7. **Strain**, ε = change in length per unit length (dimensionless)
   $$\varepsilon = \Delta L/L$$
8. **Modulus of elasticity**, E = Stress per unit strain (N/mm^2)
   $$E = \text{stress}/\text{strain} = \sigma/\varepsilon$$
   A material with a high E value is stiff and hence deforms relatively little under load.

2.8 Exercises

E2.1 *Figure 2.15* shows some traction weights (loads in kg) being applied to a patient's leg.
a) Determine the forces in the cables in newtons.
b) Find the total horizontal and vertical components of the force.
c) Determine the magnitude and direction of the resultant force.
d) If the cables are 1.5 mm diameter, what is the stress in the cable supporting the 3.5 kg mass?
a) *forces = 19.6 N and 34.3 N*
b) $\Sigma F_H = 29.7\ N \leftarrow, \Sigma F_V = 36.8\ N\uparrow$
c) *resultant = 47.3 N, direction = 51.1° to horizontal*
d) *stress = 19.4 N/mm²*

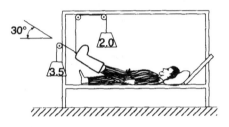

Figure 2.15
Traction weights in kg

E2.2 For the structure shown in *figure 2.16*:
a) Determine the vertical and horizontal components of the 20 kN force at C.
b) Calculate the vertical and horizontal reactions at the supports A and B.
a) $F_{CV} = 10.00\ kN\downarrow, F_{CH} = 17.32\ kN\rightarrow$
b) $R_{AV} = 14.00\ kN\downarrow, R_{AH} = 17.32\ kN\leftarrow, R_{BV} = 24.00\ kN\uparrow, R_{BH} = 0$

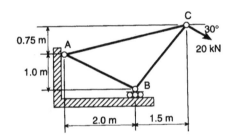

Figure 2.16
Pin-jointed structure

E2.3 For the structure shown in *figure 2.17* determine:
a) The support reactions at A and C.
b) The stress in the vertical rod.
c) The extension of the vertical steel rod in mm if the rod has a modulus of elasticity, E, of 205 000 N/mm^2.
a) $R_A = 6\ kN, R_C = 4\ kN$
b) *stress = 51 N/mm²*
c) *extension = 0. 75 mm*

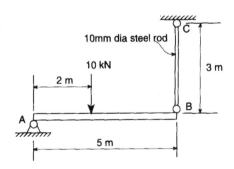

Figure 2.17
Beam and tie structure

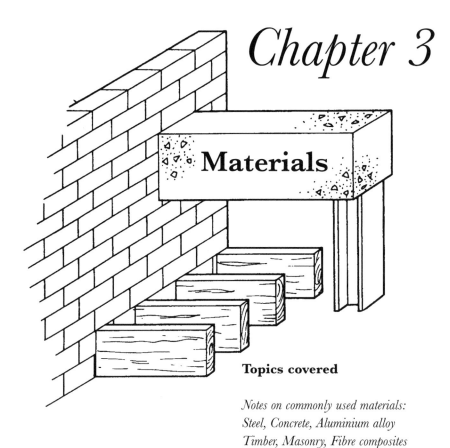

Chapter 3

Materials

Topics covered

Notes on commonly used materials:
Steel, Concrete, Aluminium alloy
Timber, Masonry, Fibre composites

Factors affecting the selection of structural materials:
Strength and stiffness, Durability
Fatigue, Brittle fracture, Creep,
Fire resistance, Weight, Economics
Environmental factors

Materials and limit state design

3.1 Introduction

Some knowledge concerning the behaviour of materials is vital if safe, reliable and long-lasting structures are to result. Materials fall into three categories. **Natural materials** such as stone and timber have been used for centuries as building materials, and their properties are well understood by craftsmen for use in small-scale building. However, because they are natural materials, they are of variable quality and often contain significant defects. This means that, for use in large-scale engineered structures, they need to be carefully selected and subject to large material safety factors to ensure safety. **Manufactured materials** such as steel and aluminium alloy are produced under carefully controlled factory conditions, with frequent testing and monitoring throughout the manufacturing process. This obviously produces a more predictable and consistent material which is reflected in lower material safety factors being required. Concrete lies somewhere between these two – being manufactured from natural materials with little intermediate processing. The third category is **new materials**, such as fibre reinforced composites. These are highly manufactured materials, but unlike steel, have not been in existence long enough to be fully understood. Because of this ignorance they tend to be used with fairly high material safety factors. As research into their properties and use advances we would expect these factors to come down.

We could also perhaps talk about a fourth category of **old materials**. Materials such as cast iron and wrought iron are now no longer used, but the designer may occasionally be faced with the problem of assessing the strength of old structures containing them. Stronger and cheaper alternatives have now been found for both of these examples, and cast iron possesses the unfortunate property of being brittle and hence unpredictable in tension.

This chapter briefly describes the commonly used materials. We then go on to examine the various factors that must be considered when selecting a material for the building of safe and durable structures. This is firstly done by considering the properties of mild steel, which is one of the most commonly used materials. The performance of the other materials is then given for comparison.

In certain cases data is provided which is referred to in later chapters of the book as the need arises.

3.2 Commonly used structural materials

The six materials that we will be considering in detail are steel, concrete, timber, masonry, aluminium and fibre composites. This covers the vast majority of materials used in modern structures.

3.2.1 *Steel*

The basic raw material for steel is iron ore which is first converted to pig iron. This contains both iron and carbon. The percentage of carbon is then reduced to produce steel. The quantity of carbon remaining, and other factors such as the rate of cooling, influence the properties of the finished steel very significantly, and so good quality control is required to produce a consistent and reliable steel for structural purposes. The commonest type is **mild steel** and this is very widely used in structural frames. It combines the qualities of being relatively strong, cheap, **ductile** (i.e. not brittle) and easy to weld.

A stronger grade known as **high yield steel** is also produced, and this is commonly used for reinforcing bars in concrete. High yield steel can also be used in structural frames but it is less likely to be available at local steel stockholders. Although it is stronger than mild steel, as we shall see later, it has the same stiffness, and consequently the deflection of structures may become more of a problem.

Other elements can also be added to produce special purpose steels. For example chromium is added to produce stainless steel; however, this tends to be too expensive for general use in structures. Very high strength steels are produced for the reinforcing wires in pre-stressed concrete. The process of pre-stressing is explained briefly in *chapter 9*.

Most steel structures use **standard structural sections**. Each major industrial country produces a range of standard cross-sectional shapes which are widely available for general use. Red-hot billets of steel are passed through a series of rollers until the shape is gradually transformed to the desired profile. These are then cut to transportable lengths for distribution throughout the country. *Figure 3.1* shows the types of standard rolled section available. Each profile is available in a range of standard sizes and weights. The different weights are achieved by moving the rollers further apart in the manufacturing process to produce thicker sections. Details of some of the available standard sections are provided in the *Appendix*. Only a small percentage of those available are included.

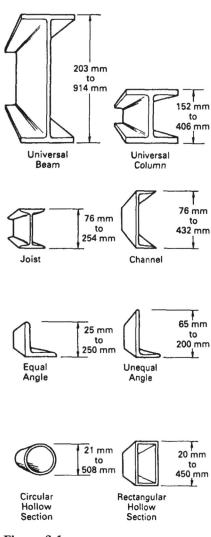

Figure 3.1
Standard rolled-steel sections

As stated above, most reinforcing bars for concrete are made of high yield steel but some mild steel bars are used. They can be distinguished visually by the fact that mild steel bars are smooth and high yield bars have an irregular surface pattern to improve bond with the concrete. See *figure 3.2*. The standard available bar diameters are shown in *table 3.1* together with their cross-sectional areas which will be used for worked examples later in this book.

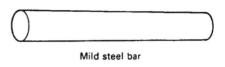

Mild steel bar

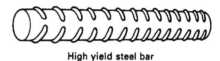

High yield steel bar

Figure 3.2
Types of concrete reinforcing bar

Table 3.1
Diameters and areas of reinforcing bars

Bar dia. (mm)	6	8	10	12	16	20	25	32	40
C/s area (mm^2)	28	50	79	113	201	314	491	804	1256

3.2.2 *Concrete*

Concrete has four principal constituents:

- cement
- fine aggregate (i.e. sand)
- coarse aggregate (i.e. chippings)
- water.

The commonest type of cement is known as **ordinary portland cement** (**OPC**). Special cements are available (such as rapid-hardening cement) and there is also an extensive range of admixtures to modify the properties of concrete. Examples are waterproofing agents and plasticisers which improve the flow of the wet concrete mix.

The proportions of the constituents are varied to produce a workable mix of the required strength. This process is known as **mix design** and is beyond the scope of this book. A general purpose mix suitable for small foundations is obtained if the ratio by weight of cement/fine aggregate/course aggregate is 1:2:4. For more important structural work the strength of a mix is defined in terms of the crushing strength of a sample cube at 28 days. Typical crushing strengths vary from 30 N/mm^2 to 50 N/mm^2. A concrete mix with a crushing strength of, say, 40 N/mm^2 is referred to as **grade C40** concrete.

When water is added to cement a chemical reaction called **hydration** takes place. The setting process should not be thought of as 'drying out', and in fact concrete can be placed quite successfully below water. Normal concrete starts to set about 45 minutes

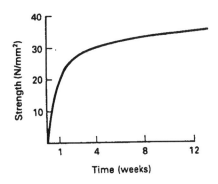

Figure 3.3
The increase in strength of concrete with time

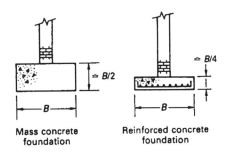

Figure 3.4
Concrete foundations

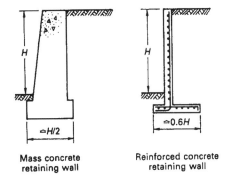

Figure 3.5
Concrete retaining walls

after the water is added, and it should not be disturbed after this **initial set** has begun. The concrete should then be protected against evaporation and excesses of temperature until the **final set** is complete after three to six days. This is known as **curing**. The concrete continues to harden for many years although the rate of increase declines rapidly after fourteen days. This is shown in *figure 3.3* for grade C30 concrete (i.e. concrete with a 28-day cube crushing strength of 30 N/mm^2). It can be seen that the 7-day strength is roughly two-thirds of the 28-day strength.

Concrete is a brittle material whose tensile strength is only about 10% of its compressive strength. For this reason it is usually reinforced with steel bars. The exception to this is the case of **mass concrete structures**. These are invariably bulky structures which rely upon the weight of the concrete to prevent significant tensile stresses arising. *Figure 3.4* shows an example of a mass concrete foundation together with its more slender reinforced concrete equivalent. *Figure 3.5* shows the same comparison for a retaining wall. Although mass concrete structures contain large volumes of concrete, they can still be economic because of the simplicity of construction.

3.2.3 *Timber*

Timber or wood can be divided into two main types – **hardwoods** and **softwoods**. Hardwoods originate from broad-leaved trees such as oak, ash and mahogany, whereas softwoods originate from coniferous trees such as spruce, pine and Douglas fir. The terms hardwood and softwood are loosely accurate, but can be misleading – balsa, as used by model aeroplane makers, is classed as a hardwood but is in fact very soft. On the other hand pitch pine, which is a softwood, is quite hard.

Softwoods tend to be quicker growing than hardwoods, and consequently are cheaper to buy. Most structures are, for this reason, manufactured in softwood. In the UK timber is currently designed to comply with BS 5268, and this is one of the few standards which is still written in terms of permissible stresses rather than the more modern limit state approach. However, Eurocode 5 uses a limit state approach.

Any timber which is used for structural purposes should be **stress graded**. Two methods of stress grading are available. It can be carried out manually by specially trained inspectors. They assess individual pieces of timber for the number and position of such defects as knots and splits, and then, if suitable, each piece is stamped with its appropriate grade. There are two manual grades

– GS (general structural) and SS (special structural). Knowing the timber species and the stress grade it is then possible to place the timber into a **strength class**, which determines the permissible stresses to be used in structural design calculations.

The second stress grading method is by machine. Each piece of timber is fed through a machine which measures the force required to bend the timber. A relationship between the bending stiffness and the appropriate stress class is then assumed, and the piece stamped accordingly.

For softwoods the strength classes range from C14 to C30, where the number refers to the ultimate bending stress in N/mm^2. For example, two of the most commonly used timbers – **whitewood** and **redwood** – are both classed as shown in *table 3.2*.

Table 3.2
Strength classifications for
whitewood and redwood timber

Stress grade	Strength class
GS	C16
SS	C24

Based on BS 5268: Part 2: 1996

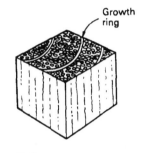

Figure 3.6
The magnified structure of timber

Figure 3.6 shows a piece of timber magnified to show its grain structure. It can clearly be seen that the tubular cells produce a highly **anisotropic** material (i.e. having different properties in different directions). This explains why, unlike steel, timber is subject to different permissible stresses depending upon whether the direction of loading is parallel or perpendicular to the grain.

Table 3.3
Standard timber sizes

Thickness (mm)	Width (mm)						
	75	100	125	150	175	200	225
38	*	*	*	*	*	*	*
47	*	*	*	*	*	*	*
63				*	*	*	*
75				*	*	*	*

Based on BS EN 336: 1995

Stress graded timber is available in a range of sizes, some of which are indicated by an asterisk in *table 3.3*. The sizes are for sawn timber, but if it is planed, to improve the surface finish, each dimension must be reduced by approximately 5 mm.

3.2.4 *Masonry*

Structural masonry covers building with bricks, stone and concrete blocks. Like concrete these are brittle materials which are weak in tension, and hence their use is restricted to situations where they largely remain in compression. An exception to this is when large wall panels are subject to horizontal wind loads. This produces bending, and a small tensile stress is often assumed to be carried by the masonry.

Bricks are usually made from clay shale which is ground up and mixed with water. They are then shaped either by being cast in moulds or by being extruded through a die (like toothpaste from a tube) and then chopped into individual bricks. The bricks are then fired in a kiln to make them hard and weatherproof. The moulded bricks often have an indentation in one or more face called a **frog**. Extruded bricks often contain holes and are known as **perforated** (see *figure 3.7*). These are chiefly to aid the distribution of heat during the firing process, but they also help to improve the bond with the mortar. Most bricks are made to a standard size which is shown in *figure 3.8a* although non-standard sizes can be made. With a standard thickness of mortar bed of 10 mm this produces a basic dimensional module of 225 mm × 75 mm as shown in *figure 3.8b*. The original clay considerably influences the strength and appearance of the finished bricks. **Facing bricks**, as the name suggests, are of good appearance but can have a very wide range of strengths. They can be used for structural work but the designer must ensure that the strength is adequate. **Engineering bricks** may not look too good, but have a minimum guaranteed crushing strength. It is important to distinguish between **bricks** and **brickwork**. Bricks are the individual units whereas brickwork refers to the complete product and includes the mortar. The crushing strength of brickwork is significantly less than that of the individual bricks.

The mortar itself is available in different grades. It commonly consists of a mixture of building sand and cement with the addition of a small amount of **plasticiser** to make the mix more workable. The proportions of cement to sand range from 1:3 to 1:8. The lower sand ratios produce a stronger mortar, but the mortar should never be stronger than the masonry units. A mortar

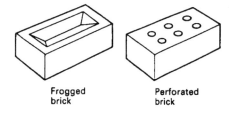

Figure 3.7
Types of brick

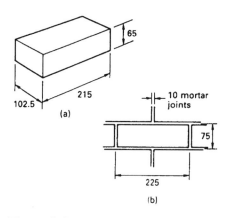

Figure 3.8
The standard brick size and module

which is too strong can cause bricks to crack under the influence of settlement or temperature movements.

Natural stone is a relatively expensive building material but is often used for prestige buildings or to blend with older structures. Again its appearance and strength are very variable, and it can be used in many forms ranging from rough random cobbles to finely sawn blocks. There are no standard sizes. The design procedures for natural stone structures are similar to those for brickwork.

Concrete blocks are one of the cheapest building materials. They are available solid or hollow (see *figure 3.9*). Special lightweight aggregates are often used in their manufacture to improve insulation properties. They are usually of poor appearance and need to be covered with plaster for internal use and a cement render for external use; however, good quality facing blocks, in various colours, are available. The commonest size of block is shown in *figure 3.10* and, when combined with a 10 mm mortar bed, this produces a basic dimensional module of 450 mm × 225 mm. Available thicknesses include 75 mm, 90 mm, 100 mm, 115 mm, 190 mm and 215 mm.

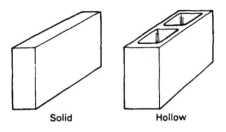

Solid   **Hollow**

Figure 3.9
Types of concrete block

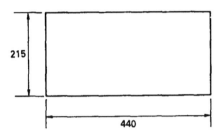

Figure 3.10
The standard concrete block size

3.2.5 *Aluminium*

Aluminium's light weight, good strength and corrosion resistance make it a viable alternative to steel in certain circumstances. Bauxite, the ore from which aluminium is obtained, is very common, but the extraction process is energy intensive which makes the finished metal relatively expensive.

Pure aluminium is too soft and weak for use as a structural material, but aluminium alloys containing about 5% of other elements such as magnesium and silicon have much improved properties. Many different alloys exist but the properties given in this book refer to a particular alloy designated as 6082 (formerly H30). This is a general-purpose structural grade with good strength, durability and machining properties. However, unlike mild steel, the heat produced when welding causes a significant reduction in strength (to about a half of its original value). This only occurs locally to the weld, but it can add considerably to the complexity of the design process. Other alloys such as 5083 (formerly N8) are less strong when unwelded but suffer much less reduction in strength when welded.

Aluminium can be extruded into complex prismatic shapes, but the available range of structural shapes is restricted to smaller sizes than the steel sections.

3.2.6 *Fibre composites*

One type of fibre composite is commonly referred to as glass fibre. Glass-fibre boats and car bodies have been commonplace for many years, but until recently their design has been based on experience and judgement rather than rational calculation. There is still no official standard document to provide design guidance, but nevertheless, more designers are using fibre composites in real structures.

Fibre-reinforced polymer composites, to give them their full title, are often referred to simply as **composites**. As the name suggests, they are formed from two distinct materials in a similar way to reinforced concrete. The fibre reinforcement is invariably glass fibre for applications in building. Carbon fibres produce about three or more times the stiffness and strength of glass, however, they cost about thirty times as much. This limits their use to high cost industries such as aerospace, formula 1 racing cars and expensive tennis rackets! In glass reinforced polymers (**GRP**) the glass fibres can be in the form of randomly oriented chopped fibres, a woven mat or unidirectional fibres – which, as the name suggests, means that most of the fibres lie in one direction. Unidirectional fibres are the most efficient for structural purposes.

The other component of a composite is the **resin polymer** or **matrix**. Three types are available – polyester, vinylester and epoxy. Polyester is the most widely used because it is cheaper than the other two, but they have better resistance to heat and chemicals.

Many techniques are available for manufacturing composites which range from early labour-intensive methods, or **hand lay-ups**, to modern processes such as **pultrusion**. This is a process whereby the fibres, in mat and unidirectional form, are pulled firstly through a bath of resin and then through a heated die which causes the resin to cure or harden. In this way long lengths of structural section can be economically produced (see *figure 3.11*). A range of standard structural sections is produced but, like aluminium, they are at the small end of the range compared to steel sections. The properties of fibre composites given in the rest of this chapter refer to pultruded sections with glass reinforcement and polyester resin.

Techniques are also now available for 'knitting' the fibres into complex shapes for the production, in moulds, of three-dimensional components.

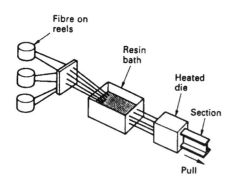

Figure 3.11

The pultrusion process for manufacturing fibre-reinforced composite structural sections

3.3 Properties of structural materials

3.3.1 *Strength and stiffness*

Strength and stiffness are probably the most important of all properties when considering whether a material is suitable for use in structures. The strength of a material obviously dominates the determination of the collapse load of a structure. Stiffness is vital to ensure that structures do not deflect too much under load: however, as we shall see later, it also affects collapse as it controls the buckling load of compression members.

The two properties are considered together because they can be investigated in the same simple test. Suppose we wished to investigate the strength and stiffness of mild steel – one of the commonest structural materials. The first step is to prepare a test specimen or **coupon**. A typical specimen is shown in *figure 3.12*. It is shaped in this way so that it will fail in the centre portion – well away from the ends which are gripped in the test machine. The specimen is then placed in a tensile testing machine which will pull it until it snaps. The load is applied in small increments and the extension over the gauge length is measured at each increment. (Modern test machines are equipped to provide a continuous readout of load and extension.)

We could now draw a graph of load versus extension for the entire test. However, this would only give us information about the particular specimen that we have tested. If we want it to provide information about the fundamental properties of the material it is preferable to change the axes of the graph:

Instead of **load** plot **stress**:

(from *chapter 2*) $\text{stress} = \dfrac{\text{load}}{\text{area}}$

Instead of **extension** plot **strain**:

(from *chapter 2*) $\text{strain} = \dfrac{\text{extension}}{\text{original length}}$

This produces the graph shown in *figure 3.13*. This plot contains some key characteristics:

- In the early stages of the test it can be seen that the plot rises steeply in a straight line. This is the **linear elastic** range.

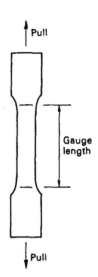

Figure 3.12
A tensile test coupon

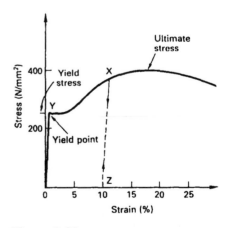

Figure 3.13
The results of a tensile test on mild steel in terms of stress and strain

Linear simply means that stress is directly proportional to strain. Elastic means that the extension is reversible – i.e. if the load was removed at this stage the specimen would return to its original length. The energy put into the steel specimen by the testing machine is stored in the specimen like in the spring of a clock.

- The tangent of the slope of the initial straight portion of the plot:

$$\tan \theta = \frac{\text{stress}}{\text{strain}}$$

We saw in *chapter 2* that this is the **modulus of elasticity**, *E*, of the material, which is of course the measure of the stiffness of the material. So the steeper the slope of the line, the stiffer the material, and hence the more resistance to deformation.

- Beyond a certain load the sample starts to stretch significantly under more or less constant load. This is the **plastic** range. If the load was now removed the specimen would not return to its original length. The specimen would feel warm if touched. This indicates that the extra energy put into the specimen has been lost in friction as the steel crystals have been dragged over each other.

- The kink in the curve between the elastic and plastic regions (point Y) is the **yield point**. The stress in the specimen at this point is the **yield stress**, σ_y. Mild steel has a very distinct yield point but, as we shall see later, many materials have an indistinct elastic–plastic transition zone.

- Beyond the flat plastic region the load starts to increase again. The sample is now **strain hardening**. This simply means that as a result of high strains the sample is getting stronger. If the load was removed from the specimen at this stage (point X) the plot would follow a straight line back to zero load (point Z). If the load was now reapplied it would retrace the line back to point X before yielding again. Strain hardening is a factory technique used to increase the yield strength of steels.

- Eventually the specimen reaches its **ultimate load** and then snaps. The failed sample would look like *figure 3.14*. The narrowing of the specimen just before failure is known as **necking**. The stress (based on the original cross-sectional area of the specimen)

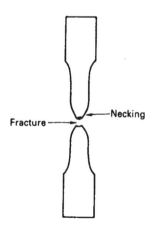

Figure 3.14
Necking of a sample at failure

is the **ultimate tensile stress**, but this is less important than the yield stress for structural design. This is because, by the time the ultimate tensile stress is reached, the structure has already suffered unacceptable permanent deformation.

- Two other points are worth defining at this stage. The **elastic limit** is the point beyond which strains are not completely reversible. The **limit of proportionality** is the point at the end of the straight-line portion of the plot. With mild steel these points effectively coincide with the yield point but this is not the case with all materials.

In general, structures must be designed so that, throughout their normal working life, stresses are always below the yield stress. To go beyond this point would imply that the structure would suffer large and permanent deformations. The plastic zone is, however, a very useful and comforting property of the material. It means that the material is **ductile** and not brittle like cast iron. Structures made from ductile materials can often avoid catastrophic collapse by deforming in such a way that loads are distributed to other supporting members. Energy is absorbed by plastic deformation. With brittle materials, on the other hand, energy is released suddenly producing violent explosive failures. Consider the difference between the failure of an aluminium drink can and a glass bottle! The permanent strain at failure is the **elongation**.

From the above it can be seen how important stress-strain characteristics are in assessing the suitability of a material. They give us information about the strength, the stiffness and the plastic properties of the material. *Figure 3.15* shows how a range of materials compare. Note that the strain has been cut off at 3% so that the elastic regions can be seen more clearly. Remember that materials such as mild steel can strain up to 25% before failure.

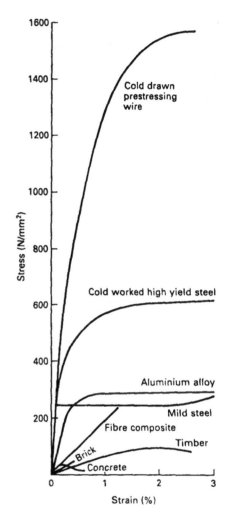

Figure 3.15
Comparative stress/strain curves for different structural materials

We can make some interesting observations:

- Note how the higher grades of steel have much superior strengths, yet the elastic modulus remains the same. This means that excessive deflection of structures may be more of a problem when using the higher grades. Also note the lack of ductility with the very high strength pre-stressing steel.
- Where a material does not have a distinct yield point, such as aluminium alloy or cold worked high yield steel, it is necessary to define a **proof stress** for use in strength calculations. This

44   Understanding structures

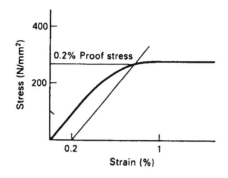

Figure 3.16
The method of obtaining a proof stress for materials without a distinct yield point

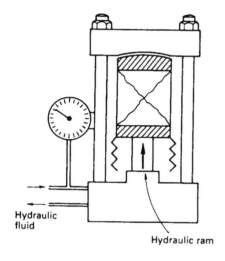

Hydraulic fluid

Hydraulic ram

Figure 3.17
A concrete cube being tested in compression

is the stress which occurs at a given permanent strain – say 0.2%. *Figure 3.16* shows how this is obtained by drawing a line parallel to the linear elastic portion of the curve.

- Concrete and masonry are normally restricted to use as compressive materials and the curves shown are based on compression tests. Concrete is tested by crushing cubes as shown in *figure 3.17*. The normal size is either 100 mm or 150 mm. The tensile strength of concrete is about 10% of its compressive strength and is usually ignored in structural calculations.

- The curve for fibre composite indicates a brittle material with no plasticity. In fact the curve for plate glass would be very similar. This is rather misleading, however, as plate glass is an homogeneous material which would shatter as a crack instantly propagated across it. With composites, on the other hand, the fact that the glass is in the form of individual fibres means that damage is localised.

From the above stress/strain curves it is possible to extract the data shown in *table 3.4* for use in design calculations.

Table 3.4
Strength and stiffness properties of commonly used structural materials

Material	Yield stress or 0.2% proof stress (N/mm²)	Ultimate stress (N/mm²)	Permissible stress (N/mm²)	Modulus of elasticity (N/mm²)
Mild steel	275	–	–	205 000
High yield steel	460	–	–	200 000
Pre-stressing wire	–	1570	–	200 000
Concrete	–	25–60	–	28 000
Timber:				
softwood	–	16	7.5	7 000
hardwood	–	30	14	12 000
Engineering brickwork	–	15	–	20 000
Aluminium alloy	255	–	–	70 000
Glass-fibre composite	–	250	–	20 000

Example 3.1

A specimen of aluminium alloy has a gauge length of 50 mm and a cross-section which measures 20 mm × 5 mm. It gave the following results when subjected to a tensile test. Plot the stress/strain curve for the first 1.2% of strain and determine:

a. The 0.2% proof stress for the material.
b. The modulus of elasticity, E.
c. The ultimate tensile stress.
d. The percentage elongation at failure.

Load (kN)	Extension (mm)
4	0.028
8	0.057
12	0.083
16	0.115
20	0.160
24	0.221
28	0.332
29	0.395
29.5	0.652
30.9 (max.)	7.25 (at failure)

Solution

The first stage is to convert the loads into stresses and the extensions into strains:

$$\text{Stress} = \frac{\text{load}}{\text{area}} = \frac{\text{load}}{20 \times 5}$$

$$\% \text{ Strain} = \frac{\text{extension} \times 100\%}{\text{gauge length}}$$

$$= \frac{\text{extension} \times 100\%}{50}$$

The final extension of 7.25 mm is way beyond the 1.2% strain limit and so is ignored at this stage.

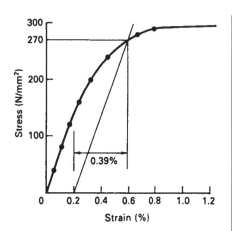

Load (kN)	Stress (N/mm²)	% Strain
4	40	0.056
8	80	0.114
12	120	0.166
16	160	0.230
20	200	0.320
24	240	0.442
28	280	0.662
29	290	0.790
29.5	295	1.304

Figure 3.18
The solution to example 3.1

This produces the curve shown in *figure 3.18*.

a. **Answer**   From the curve **0.2% proof stress = 270 N/mm²**

Modulus of elasticity, E, is given by the tangent of the slope of the straight portion of the curve.

b. **Answer**

$$\textbf{Modulus of elasticity, } \boldsymbol{E} = \frac{270}{0.0039} = \textbf{69 200 N/mm}^2$$

Finally from the values at failure:

c. **Answer**

$$\textbf{Ultimate stress, } \boldsymbol{\sigma_u} = \frac{1000 \times 30.9}{20 \times 5} = \textbf{309 N/mm}^2$$

d. **Answer**

$$\textbf{\% Elongation at failure} = \frac{7.25}{50} \times 100\% = \textbf{14.5\%}$$

3.3.2 *Durability*

Clearly when selecting a structural material it is important that it remains intact and fully functional for a reasonable lifetime.

Durability is closely related to the environment and climate in which the material is expected to perform. Thus unfired mud bricks may be adequate for small structures in very hot and dry climates. They would, however, be useless in cold wet climates, where they would disintegrate quickly with the action of rain and frost.

Factors which can affect durability are:

- Temperature
- Humidity
- Rainfall
- Degree of exposure
- Polluted industrial atmospheres
- Salty coastal atmospheres
- Abrasion from wind-borne sand
- Electrolytic contact with dissimilar materials.

Where any of these are known to be a particular problem, special care should be taken in the selection and protection of materials.

Many materials, such as steel and timber, require special treatment to increase their resistance to decay. The following brief notes describe how durability problems can be avoided for each of the common materials.

Steel structures need protection from rust, which forms on the surface of the metal in the presence of oxygen and moisture. Rusting is a progressive process which can completely reduce sound metal to a heap of rust particles (as owners of old cars will know). Most of the methods available to protect steel involve applying a protective coating. This can take the form of **galvanising** and/or a properly designed paint system. Galvanising is a process whereby a zinc coating is added. Paint systems must include compatible primer, undercoats and finishing coats. In both cases the surface of the steel must be adequately prepared to accept the coating. This can be a particular problem with rolled steel sections, as the surface is covered with loose steel flakes or millscale. This is best removed by shotblasting before the primer is applied. Paint requires routine maintenance to keep it effective, and the cost of this should be taken into account when choosing steel for use in exposed situations.

Concrete structures can be very durable and require no maintenance, but problems can arise. In reinforced concrete the concrete provides protection to the steel reinforcement, but this depends on good quality concrete and adequate **cover**. Cover is defined as the minimum distance from the surface of a reinforcing bar to the face of the concrete (see *figure 3.19*). *Table 3.5* gives the recommended minimum cover for different degrees of exposure.

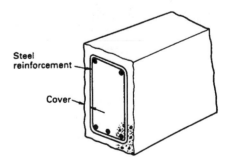

Figure 3.19
Concrete cover to steel reinforcement

Table 3.5

Durability cover to concrete reinforcement (mm)

Degree of exposure	Concrete grade				
	C30	C35	C40	C45	C50
Mild	25	20	20	20	20
Moderate	–	35	30	25	20
Severe	–	–	40	30	25
Very severe	–	–	50	40	30
Extreme	–	–	–	60	50

Based on BS 8110: Part 1: 1997

If the step reinforcement does start to corrode it increases in volume, which can cause the surface of the concrete to spall away, thus exposing the reinforcement even more.

Concrete is also susceptible to chemical attack. Excess sulphates in groundwater can attack concrete in foundations. A sample of groundwater should always be analysed as part of the initial site investigation and, if it is a problem, special **sulphate-resisting cement** can be specified. Concrete has also been found to deteriorate as a result of using certain aggregates and additives. Even road salt added to highways to prevent icing has caused problems with concrete bridges.

Aluminium alloys particularly deteriorate in the salty atmospheres of coastal regions where the surface can become pitted and covered with a white furry deposit. To overcome this they can be painted flake steel, or alternatively a very good quality finish can be obtained by **anodizing**. This is an electrolytic process which produces an inert coating on the surface of the alloy. This provides adequate protection even for yacht masts.

Aluminium alloys can also corrode if brought into direct contact with dissimilar metals. Insulating plastic washers must be used to prevent this electrolytic corrosion taking place (see *figure 3.20*).

Timber is subject to decay from both fungal attack and wood-boring insects. **Wet rot** is a fungal attack which occurs when the wood is actually wet, whereas the more sinister **dry rot** occurs when the moisture content is above 20%. Dry rot is particularly troublesome because it can spread rapidly, even through brickwork, by releasing long tendrils. It particularly thrives in damp, dark and unventilated spaces.

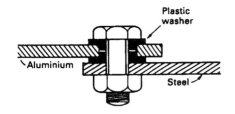

Figure 3.20

The insulation of aluminium in contact with steel to prevent electrolytic corrosion

The principal insects which cause damage are wood worm and death watch beetle. Given time, all of these infestations can reduce the strength of timber to zero.

Timber can be protected from attack by impregnating it with preservative – preferably under pressure. In addition the traditional method of protecting external timber is by coating with paint or varnish. However, the natural tendency of timber to shrink and swell with the seasons means that the coating is prone to split and flake. Also moisture can be trapped beneath the paint skin. Modern coatings are now available which are both **elastic** and **microporous**. Micro-porous coatings provide a barrier to water particles but allow water vapour to evaporate from the timber.

Brick and **stone** are generally highly durable materials as evidenced by the large number of ancient buildings which still survive. There is, however, considerable variation between different types of masonry. As usual, the main enemy of durability is water. Some types of brick and stone are particularly susceptible to frost damage. Repeated wetting and freezing can cause cracks to form and the surface of walls to flake away. Damage can also be caused by **acid rain**, which occurs when sulphur dioxide, from the burning of fossil fuels, reacts with water to form sulphuric acid.

Special masonry paints are available to provide protection, but their periodic replacement adds to maintenance costs. Sealants can be sprayed onto the surface of masonry to prevent water penetration, but the natural passage of water vapour out of the building can be impaired. The best solution is to leave masonry untreated unless a particular problem arises.

Fibre composites are highly durable and require no maintenance. For this reason they are used in hostile chemical and atmospheric environments. Vinylester and epoxy resins are the most resistant.

3.3.3 Fatigue

Consider the paperclip shown in *figure 3.21*. If it is bent beyond its yield point and then back again, only once, it will not suffer any reduction in strength. (In fact the material may have increased in strength owing to strain hardening.) If, however, this is repeated several times it will break. This is a **fatigue** failure and is caused by **fluctuations** in stress. If the paperclip had been bent gently

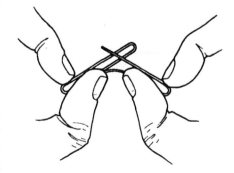

Figure 3.21
A paperclip undergoing a fatigue test

so that it stayed within its elastic range it could still suffer a fatigue failure, but the number of cycles required would be very much greater – say a million! Fatigue failure is caused by small cracks which gradually propagate through the material with each stress cycle.

Fatigue is therefore a problem in structural members which are subject to variations in stress, and this applies to all structures subject to varying loads. For normal structural members in buildings, however, the variation in stress and the number of cycles are not great enough to cause a problem. Consequently it is rare to have to check such structures for fatigue.

It is a problem in vehicles which are subject to vibration and dynamic loads, and has been responsible for the failure of chassis members and aircraft wings. It can often explain why engine components eventually break in old cars. Special consideration must be given to fatigue in the following types of structure:

- Cranes and hoists
- Lightweight structures subject to flutter or oscillations in wind
- Bracing members subject to frequent stress reversals.

Figure 3.22 shows the relationship between stress fluctuation (S) and the log of the number of cycles (N) to cause failure for two mild steel members. One is a simple member with smooth curves and carefully ground flush connections. The other is a member with sudden changes in cross-section, holes and crudely welded attachments. It can be seen that the latter member deteriorates much more rapidly from fatigue. This is because the abrupt changes in cross-section produce **local stress concentrations** – i.e. points of high stress which act as triggers for the propagation of fatigue cracks. Corrosion can cause pitting of the surface of metals which also induces early fatigue failure. Note that the levelling-off of the S–N curves indicates that provided stress amplitudes are kept below a certain **fatigue limit** then an infinite life can be expected. This is not the case with all materials, and *figure 3.23* shows the S–N curves for aluminium alloys and glass-fibre composites, in which it can be seen that the fatigue strength continues to reduce indefinitely.

Most materials suffer from fatigue and this eventually leads to a stress limit of around 20–70% of the static ultimate stress. The usual way of dealing with fatigue is to reduce the stress range in the material by increasing the size of structural members, and also to avoid unnecessary local stress concentrations.

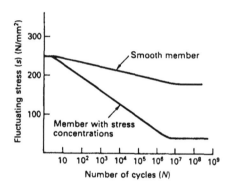

Figure 3.22
Fatigue failure curves for mild steel

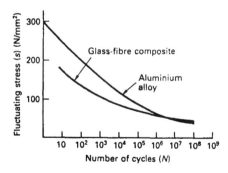

Figure 3.23
Fatigue failure curves for aluminium alloy and glass-fibre composite

3.3.4 *Briffle fracture*

Under certain conditions normally ductile mild steel and some plastics can become brittle. The three conditions which are most influential are impact loading, low temperatures and relatively thick material. Ships have been known to break in two as a result of brittle fracture in Arctic storm conditions.

If structures are thought to be at risk from brittle fracture, special consideration must be given to the selection of the material. For example, special grades of steel are available which retain their ductility at lower temperatures. Brittle fracture is not a problem with the other structural materials.

3.3.5 *Creep*

When a load is applied to a structural member it causes an immediate strain which produces deformation. If the load is left in place it continues to deform slowly with time. This time-dependent strain is known as **creep**, and materials which exhibit it are **visco-elastic**. The roofs of old buildings sometimes sag because of creep deformation of the roof timbers. Some materials, such as bitumen, are capable of enormous creep strains under small stresses. These highly **visco-elastic** materials are not generally suitable for use in structural members.

In **concrete** the amount of creep strain is proportional to stress. *Figure 3.24* shows the expected amount of creep strain with time for each N/mm^2 of stress.

Steel is not generally affected by creep at room temperature but it becomes significant at higher temperatures, and so it is particularly a problem in pressure vessels or other production plant which gets hot. A creep phenomenon which is significant at room temperature is **relaxation**. If steel is held at a high stress under constant strain (i.e. the length is fixed) the stress will reduce with time. *Figure 3.25* shows a typical relaxation curve for pre-stressing steel. The combination of concrete creep and steel relaxation means that some of the stressing force in pre-stressed concrete cables is lost. This must be allowed for in calculations.

Timber creeps very little at stress levels below 70% of ultimate, but above this figure it increases rapidly. However, with properly designed structures the stresses should be well below this level.

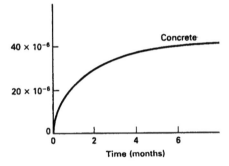

Figure 3.24
A creep curve for concrete

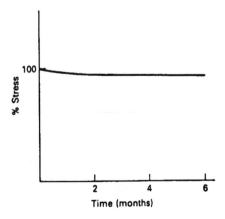

Figure 3.25
Stress relaxation of steel due to creep at constant strain

For **glass-fibre composites**, at normal stress levels, creep can be expected to add about 25% to the immediate elastic strains.

Bricks are not affected by creep; however, dry bricks, straight from the kiln, are subject to swelling when exposed to a damp atmosphere.

Beams are often constructed with a slight upward camber to allow for the fact that in the long term creep may cause some downward deflection.

3.3.6 *Fire resistance*

If materials are to be suitable for use in building structures they must satisfy certain requirements when subjected to fire. The ideal material would:

- Be incombustible and hence not spread flames
- Not produce smoke or fumes
- Retain its strength as it gets hot.

Surprisingly combustibility, or the ability to burn, is not necessarily regarded as the principal problem. More important is the ability to retain structural strength for a period of time in order to allow people to escape. For this reason structural members are required to resist fire for a specified period depending on the use of the building. This period ranges from half an hour for small low rise buildings to four hours for the basements of large multi-storey buildings. The provision of adequate means of escape and the division of buildings into fireproof compartments are also very important factors but beyond the scope of this book.

In a typical fire the temperature can rise to 1200°C and be at 800°C within just half an hour. Most materials suffer a reduction in both strength and stiffness as temperature rises. *Figure 3.26* shows the effect of temperature on the original strength of structural materials. It can be seen that all our structural materials will be effectively destroyed. The key question is how long will this take?

Metals suffer from the disadvantage that they are good thermal conductors, and so the heat spreads throughout the structural members very quickly. For this reason all metal structural members which support floors and walls must be protected by

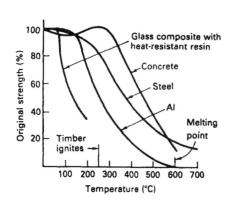

Figure 3.26

The variation in strength of structural materials with temperature special fireproof cladding. This can add considerably to the cost of steelframed buildings.

Concrete and **masonry** transmit heat roughly forty times more slowly than metals. Also because of their relative lack of strength, compared to metals, the structural members are more bulky. This means that these materials perform well in fires. Any steel reinforcement must be protected by adequate concrete cover, and national standards specify minimum values for different types of structural member. Concrete made with aggregates which contain silica can suffer from explosive spalling of the exposed concrete surface, but this process should be slow enough to allow escape.

Timber is the only truly combustible material in our list, however, it transmits heat even more slowly than concrete. This means that although the outside of a structural member may be charred, the inside still retains its strength. Timber typically chars at the rate of about 0.6 mm/minute. We can therefore provide a half-hour fire resistance by simply adding 20 mm to each face of structural members. Special paints are available to reduce the spread of flame over the surface of timber.

Fibre composites can be manufactured with special heat-resisting resins to improve their performance in fire. Care needs to be taken to select resins which will not give off toxic fumes. No firm guidelines yet exist for the protection of structural members made from fibre composites, and until they do it is prudent to treat them in the same way as steel members.

3.3.7 *Weight*

As we shall see in the next chapter, the weight of a building is usually greater than its contents. This means that structural members can be made significantly smaller and cheaper if the structure itself is made lighter. On the other hand weight can often be useful when resisting wind loads. The unit weights of structural materials are given in *table 4.1*. It is interesting to compare the strength of the different materials per unit weight – known as **specific strength**. *Figure 3.27* shows the relative specific strengths. With many structures the design is limited by excessive deflections rather than strength. In this case the stiffness per unit weight, i.e. **specific modulus**, is important and this is compared in *figure 3.28*.

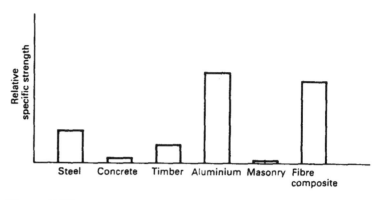

Figure 3.27
Design strength per unit weight for structural materials

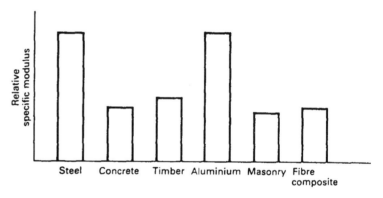

Figure 3.28
Modulus of elasticity per unit weight for structural materials

3.3.8 Cost

Apart from a few prestigious status symbols, structures are designed so that they will perform their function as cheaply as possible. Relative cost comparisons between materials are complex and constantly changing as a result of technological developments and changes in world markets. A crude measure of structural economy can be obtained by comparing structural strength per unit cost (see *figure 3.29*). These comparisons are obtained by dividing the typical design strength of each material by the cost of a unit volume of that material. Labour costs are ignored and no account is taken of the fact that some materials can be made into more efficient shapes, or that some materials may require fire protection etc. It should also be remembered that concrete and masonry can only be used in compression. All of the

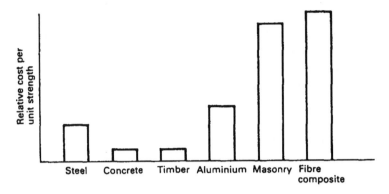

Figure 3.29
Relative cost of structural materials per unit of stress carried

materials considered here are economically viable under certain circumstances.

3.3.9 *Environmental factors*

There is a growing awareness of the importance of environmental issues in construction. Firstly the designer should ensure that the workers involved in the actual construction are not put at risk by handling hazardous materials. As an example, asbestos was formerly used extensively in roof sheeting and fireproofing materials, however, it has now been established that inhalation of asbestos dust can cause cancer. Problems have arisen from the fumes given off when welding certain materials in unventilated spaces. Fumes emitted by a particular timber preservative have caused illness in both workers and subsequent building users. In all cases care must be taken to comply with current health and safety requirements.

Designers may also wish to take account of wider environmental concerns when selecting materials. In order to protect the rainforests, the use of tropical hardwoods should be restricted to those obtained from properly managed schemes where timber is replaced. The long-term re-use of materials after demolition of structures may also be taken into account. Steel is easily recycled and concrete and masonry can be used as hardcore. Fibre composites may present a problem here as they are not biodegradable, however, they are not yet used widely enough to present a problem.

Only timber can be regarded as a renewable resource, and the manufacture of all the other structural materials uses up limited natural resources and energy. The cost of purchasing a material can, however, be a good crude measure of its environmental importance; i.e. if it is cheap, then its raw materials are probably available in abundance and its manufacture does not consume a great deal of energy. Unfortunately this is not foolproof, and wider environmental factors may not be reflected in the purchase price.

3.4 Materials and limit state design

3.4.1 *Characteristic strength*

Table 3.4 gave typical values of stress for different structural materials. Clearly they are based on strength tests carried out on, not one, but many specimens. No two tests would give exactly the same result, so how is a single strength value arrived at? An obvious answer would be to take the mean or average test value. This would not, however, be satisfactory as, to ensure adequate safety, we really need to concern ourselves more with the weaker specimens. Limit state codes define the **characteristic strength** of a particular grade of material as:

'The strength, below which, not more than 5% (or one in twenty) samples will fail.'

Experience has shown that, in general, the strength of a large number of samples of the same structural material will be distributed about the mean value in the form of a **normal distribution**. If we plot strength against the frequency of a particular strength occurring we get the familiar bell-shaped curve shown in *figure 3.30*. This indicates that most sample strengths are close to the mean value, but a few are significantly stronger and a few are significantly weaker. Consistent materials produce a bell with a narrow base whereas unreliable materials have a large spread. In statistical terms the amount of variation from the mean can be measured in terms of the **standard deviation**.

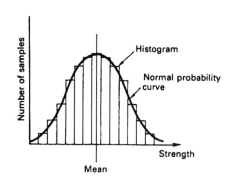

Figure 3.30
A normal distribution about the mean

$$\text{Standard deviation} = \sqrt{[\Sigma\,(x - x_{\mathrm{m}})^2/(n - 1)]}$$

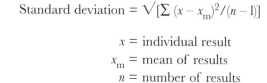

Where

$x =$ individual result
$x_{\mathrm{m}} =$ mean of results
$n =$ number of results

It can be shown from the mathematics of the normal distribution curve that:

Characteristic strength = mean value − 1.64 × standard deviation

Using the above formula should ensure that not more than 5% of specimens will fall below the characteristic strength.

3.4.2 Partial safety factor for strength, γ_m

In critical state design, the characteristic strength of the various structural materials must be divided by a safety factor before it can be used in design calculations. This is known as a **partial safety factor** because other safety factors are applied to the loads on the structure. The size of the load factor depends upon the consistency, reliability and state of knowledge concerning the material. *Table 3.6* indicates typical values of safety factor:

$$\text{Design strength} = \frac{\text{Characteristic strength}}{\gamma_m}$$

Table 3.6
Partial safety factors for material strength, γ_m

Material	Safety factor γ_m	Source
Structural steelwork	1.0	BS 5950
Steel reinforcement	1.15	BS 8110
Concrete	1.5	BS 8110
Timber	1.3*	Eurocode 5
Masonry	2.5–3.5**	BS 5628
Aluminium	1.2	BS 8118
Fibre composite	1.7	author

*Modified to take account of the duration of the load.
**Depending upon the degree of quality control.

Example 3.2

Ten concrete cubes were prepared and tested by crushing in compression at 28 days. The following crushing strengths in N/mm² were obtained:

44.5   47.3   42.1   39.6   47.3   46.7   43.8   49.7   45.2   42.7

Determine:
a. the mean strength
b. the standard deviation
c. the characteristic strength
d. the design strength.

Solution

result (x)	$(x - x_m)$	$(x - x_m)^2$
44.5	−0.4	0.2
47.3	2.4	5.8
42.1	−2.8	7.8
39.6	−5.3	28.1
47.3	2.4	5.8
46.7	1.8	3.2
43.8	−1.1	1.2
49.7	4.8	23.0
45.2	0.3	0.1
42.7	−2.2	4.8
448.9		80.0

a. Answer

$$\textbf{Mean strength } x_m = \frac{448.9}{10} = \textbf{44.9 N/mm}^2$$

b. Answer

$$\textbf{Standard deviation} = \sqrt{[(x - x_m)^2/(n - 1)]} = \sqrt{(80/9)}$$
$$= \textbf{2.98 N/mm}^2$$

c. Answer

$$\textbf{Characteristic strength} = 44.9 - (1.64 \times 2.98)$$
$$= \textbf{40.0 N/mm}^2$$

d. Answer

$$\textbf{Design strength} = \frac{40.0}{\gamma_m} = \frac{40.0}{1.5*}$$
$$= \textbf{26.7 N/mm}^2$$

*value obtained from *table 3.6*

> *Comment – In practice the calculation of a reliable standard deviation should be based on many more test results – say about a hundred. A computer or calculator with statistical functions then becomes indispensable.*

3.5 Summary of key points from chapter 3

1. The principal structural materials are **steel**, **concrete**, **timber**, **masonry** (bricks and blocks), **aluminium** and more recently **fibre-reinforced polymer composites**.
2. Structural steel work is largely fabricated from standard structural sections made from **mild steel**, whereas the bulk of steel used for concrete reinforcement is **high yield steel**.
3. Concrete is made from cement, fine aggregate (sand), coarse aggregate (gravel) and water. It is weak in tension and so is usually reinforced with steel bars.
4. The most important properties of structural materials are **strength** and **stiffness**. These are determined by testing specimens of the material in a testing machine and producing stress/strain curves.
5. The curve for mild steel is typified by a **linear elastic** region (slope = **elastic modulus**, E) and, following the **yield point**, a **plastic** region where the strains are large and non-reversible.
6. Durability is also an important property and steel and timber usually require special protection. With reinforced concrete structures, the steel reinforcement must be protected by an adequate thickness of concrete known as **cover**.
7. Structural steelwork often requires special protection from fire.
8. In limit state design the **characteristic strength** is defined as that strength, below which, not more than 5% of tested samples will fail. For a series of test specimens:

Characteristic strength = mean value – 1.64 × standard deviation

9. The characteristic strength must then be divided by a **partial safety factor for material strength**, γ_m, before it becomes a **design strength** for use in structural calculations.

3.6 Exercises

E3.1 Select suitable structural materials for the following applications. Justify your decision by referring to the properties that make your choice suitable.

a) An underground water-storage reservoir.
b) A short-span foot-bridge in a corrosive chemical plant.
c) A portable grandstand structure for one-off sporting events.

E3.2 A tensile stress/strain test is carried out on a sample of mild steel:

a) What are the two most important structural parameters that can be obtained from the test?
b) What is meant by the 'gauge length' of a specimen?
c) Why is a certain amount of plastic behaviour a good thing for a structural material?
d) Why is the ultimate stress of less interest to designers than the yield stress?
e) What three points effectively coincide in the stress/strain curve for mild steel?

E3.3 A circular mild steel specimen was subjected to a tensile test. The results for the early part of the test are shown below. Plot the stress/strain curve and determine:

a) The yield stress, σ_y.
b) The modulus of elasticity, E.

diameter of specimen = 5.0 mm
gauge length   = 40 mm

Load (kN)	Length (mm)
0.00	40.00
1.00	40.01
2.00	40.02
3.00	40.03
4.00	40.04
5.00	40.05
5.32	4.006
5.38	40.07
5.34	40.08

a) *yield stress,* $\sigma_y = 273 \, N/mm^2$
b) *modulus of elasticity,* $E = 204\,000 \, N/mm^2$

E3.4 Six samples of steel reinforcing bar were tested and gave the following yield stresses in N/mm^2. Determine the design strength to be used in reinforced concrete calculations.

455   467   449   452   471   462

design strength $= 387 \, N/mm^2$

Chapter 4

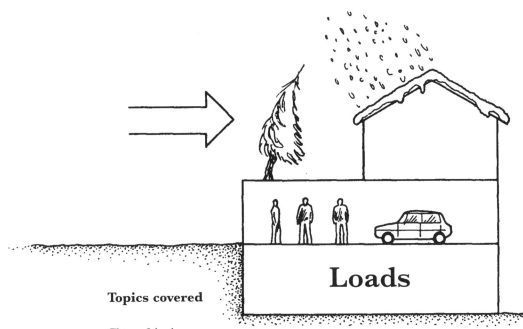

Loads

Topics covered

Types of load:
Dead load, Imposed floor load
Imposed roof load, Wind load
Loads from liquids and soils

Loads and limit state design
Calculation of loads

4.1 Introduction

We have already defined the purpose of a structure as being to transmit **loads** from one place to another. This chapter examines the various types of load that can act on a structure, and then goes on to explain how the loads carried by individual structural members are calculated. This vital stage in the design process can be quite tricky – particularly if it is necessary to analyse an old existing structure.

There are three principal types of load that must be considered:

- **Dead** loads
- **Imposed** loads
- **Wind** loads.

It is also sometimes necessary to take account of loads which do not fit neatly into the above categories, such as those from liquids and soil pressures.

In addition to the above **primary** loads, structures can be subjected to **secondary** loads from temperature changes, shrinkage of members and settlement of supports. However, for the simple structures discussed in this book these are rarely a problem, and hence will be ignored.

Eurocodes use a different terminology with respect to loads. They refer to loads as **actions**. Dead loads are **permanent actions** and imposed loads are **variable actions**. A category of **accidental actions** is also introduced, which covers such things as gas explosions and impact from vehicles. Specific values given for loads in Eurocodes vary somewhat from the values provided in this book which are based on British Standards; however, the underlying principles are similar.

4.2 Dead loads

Dead loads, as the name suggests, are those loads on a structure which are permanent and stationary. The actual self-weight of the structural members themselves normally provides the biggest contribution to dead load. Permanent non-structural elements such as tiled floor finishes, suspended ceilings, service pipes and light fittings must also be included if significant. Obviously, the dead load arising from structural members depends upon the density of materials used. Rather than express densities in terms

Table 4.1
Unit weights of various building materials (kN/m^3)

Aluminium	24
Bricks	22*
Concrete	24
Concrete blocks (lightweight)	12*
Concrete blocks (dense)	22*
Glass-fibre composite	18
Steel	70
Timber	6*

*Subject to considerable variation.

of kg/m^3, *table 4.1* gives **unit weights** in terms of kN/m^3. This means that we do not need to worry about converting mass to force. Items of permanent plant such as lift machinery and air conditioning equipment are also considered to be dead loads; however, most machinery and computers are assumed to be movable and hence **not** dead load.

It is necessary to calculate the dead loads of structural members using the above unit weights. As we shall see later, when we come to evaluate bending moments in beams, it is often more convenient to express the dead load as a weight per unit length. This is simply the weight of a one metre length of member. We therefore simply calculate the volume of a piece of member one metre long and multiply it by the material unit weight. The volume of such a piece will of course be equal to the cross-sectional area of the member in square metres (multiplied by the length in metres which is one).

To get the total dead weight of the member we can of course simply multiply the weight per unit length by the length of the member in metres. It is vital that the correct units are used for dead weight – kN/m indicates a weight per unit length and kN indicates a total weight. This is illustrated in the following example.

Example 4.1

Figure 4.1 shows a precast concrete beam which is 10.5 m long.

a. Calculate the weight of the beam per unit length in kN/m.
b. Calculate the total weight of the beam.

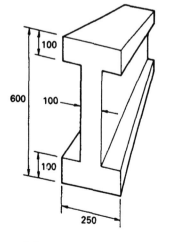

Figure 4.1
A precast concrete beam

Solution

a.

$$\text{Cross-sectional area of beam} = (0.6 \times 0.25) - (0.4 \times 0.15)$$
$$= 0.09 \text{ m}^2$$

From *table 4.1*

$$\text{Unit weight of concrete} = 24 \text{ kN/m}^3$$

Answer

Weight per unit length $= 0.09 \times 24 = $ **2.16 kN/m**

b. **Answer**

Total weight of beam $= 2.16 \times 10.5 = $ **22.68 kN**

In addition to the above it is usually more convenient for the weights of certain sheet materials to be quoted in terms of kN/m^2 and these are provided in *table 4.2*.

Table 4.2
Unit weights of various sheet materials (kN/m^2)

Acoustic ceiling tiles	0.1
Asphalt (19 mm)	0.45
Aluminium roof sheeting	0.04
Glass (single glazing)	0.1
Plaster (per face of wall)	0.3
Plasterboard and skim	0.15
Rafters, battens and felt	0.14
Sand/cement screed (25 mm)	0.6
Slates	0.6
Steel roof sheeting	0.15
Timber floorboards	0.15
Vinyl tiles	0.05

The dead load of a floor or a roof is generally evaluated for one square metre of floor or roof area. This is illustrated in the following example.

Example 4.2

The floor in a multi-storey office building consists of the following:

Vinyl tiles
40 mm sand/cement screed
125 mm reinforced concrete slab
Acoustic tile suspended ceiling

Determine the dead load in kN/m².

Solution

A cross-section through the floor is shown in *figure 4.2*.

table 4.2	vinyl tiles	= 0.05
table 4.2	screed	= 0.6 × 40/25 = 0.96
table 4.1	concrete slab	= 0.125 × 24 = 3.00
table 4.2	acoustic tiles	= 0.10
Answer	**dead load/m² =**	**4.11 kN/m²**

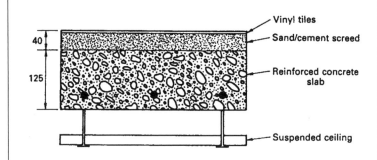

Figure 4.2
Floor construction in an office building

The dead load of a wall acting on a beam is generally given in terms of the load per metre-run along the beam. In other words, what is the weight acting on each metre length of beam?

Example 4.3

Figure 4.3a shows the outer wall of a multi-storey building which is supported on a beam at each floor level. The walls consist of a 1.2 m height of cavity wall supporting 1.3 m high double glazing. The cavity wall construction is 102.5 mm of brickwork, a 75 mm cavity and 100 mm of plastered lightweight concrete blockwork.

Determine the dead load on one beam in kN/m of beam.

Solution

Each metre length of beam supports the dead load shown in *figure 4.3b*.

table 4.1	brickwork =	$1.2 \times 0.1025 \times 22$ =	2.71
table 4.1	blockwork =	$1.2 \times 0.1 \times 12$ =	1.44
table 4.2	plaster =	1.2×0.3 =	0.36
table 4.2	double glazing =	$2 \times 1.3 \times 0.1$ =	0.26
Answer load on beam =			**4.77 kN/m**

4.3 Imposed loads

Imposed loads (or live loads) are those movable loads which act on a structure as it is being used for its designed purpose. This includes such things as people, furniture, cars, computers, machinery and cans of beans. Imposed loads are subdivided into two categories: **imposed floor loads** and **imposed roof loads**.

4.3.1 *Imposed floor loads (BS 6399: Part 1)*

Most buildings have a working life of at least fifty years and it is clearly an impossible task to predict with any accuracy the loads which will be placed on the floors throughout that period. National standards exist which attempt to give reasonable upper estimates to the loads that can occur on the floors of buildings used for different purposes. If, during the life of a building, its purpose is changed, say from a private house to a museum, it is usually necessary to carry out calculations to prove that the floors are strong enough to carry the new loads. Some typical floor loads are given in *table 4.3*.

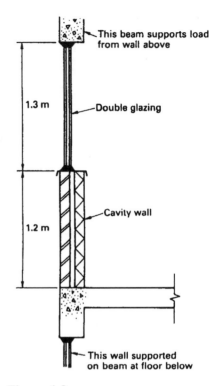

Figure 4.3a
Cross-section through the wall of a multi-storey building

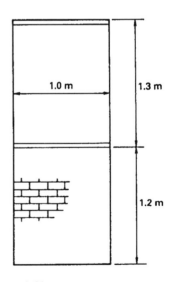

Figure 4.3b
Load on each metre of a beam

Table 4.3
Floor loads in kN/m^2

Art galleries	4.0
Banking halls	3.0
Bars	5.0
Car parks	2.5
Classrooms	3.0
Churches	3.0
Computer rooms	3.5
Dance halls	5.0
Factory workshop	5.0
Foundries	20.0
Hotel bedrooms	2.0
Museums	4.0
Offices (general)	2.5
Offices (filing)	5.0
Private houses	1.5
Shops	4.0
Theatres (fixed seats)	4.0

Based on BS 6399: Part 1

The above loads are assumed to occur over the entire floor area. There is an alternative requirement for a concentrated load to be applied at any point on the floor, but this is rarely critical in structural calculations.

Example 4.4

The average weight of a person can be assumed to be 80 kg.

a. *Table 4.3* indicates that a bar should be designed for a load of 5 kN/m^2. Calculate how many people that represents standing on one square metre of floor.

b. A student has a flat in a private house. If the living room measures 4.0 m × 5.5 m, how many people can be invited to a party if the floor loading is not to be exceeded?

Solution

a. Force exerted by one person = $80 \times 9.81 = 785$ N

Answer

$$\textbf{Number of people in one m}^2 = \frac{5 \times 10^3}{785} = \textbf{6.4}$$

b.   Floor area = $4 \times 5.5 \ = 22$ m^2

From *table 4.3* load for domestic houses = 1.5 kN/m^2

Answer   **Number of people** $= \dfrac{22 \times 1.5 \times 10^3}{785} = \textbf{42}$

Warning! – *Before you rush out to invite 42 people to your party, the above loads are assumed to be static. Certain types of dancing can cause dynamic effects which greatly increase the effect of the load on the structure. This is why soldiers are ordered to break step when marching over slender bridges. Also the above calculation assumes that the people are uniformly distributed throughout the room. If people are forced towards the centre of the room because of furniture around the perimeter, this will also increase the stress in the structure.*

4.3.2 Imposed roof loads (BS 6399: Part 3)

If access is provided to a roof, then it must be designed to support the same imposed load as the adjacent floor area. This implies a minimum of 1.5 kN/m^2. Where access is only for the purpose of maintenance, the predominant load on roofs in most countries is caused by a build-up of snow. The magnitude of load to be considered depends upon several factors including:

- Geographical location
- Height above sea level
- Shape of roof
- The wind redistributing the snow in drifts.

In the UK the standard gives basic snow loads which range from 0.3 kN/M^2 on the south coast to 1.0 kN/m^2 in northern Scotland. However, for simple flat roofs it is traditional to use a value of 0.75 kN/m^2.

For obvious reasons, less snow will form on steeply pitched roofs. The standard therefore allows us to reduce the intensity of snow loading on roofs where the pitch exceeds 30°. For a roof of slope $a°$:

the basic snow load is multiplied by $(60 - a)/30$.

When the roof pitch is 60° or more the snow load is zero.

Example 4.5

Figure 4.4a shows the plan and front elevation of a house roof which is formed by a series of 7 m wide timber trussed rafters spaced at 0.6 m. The basic snow load can be assumed to be 0.75 kN/m².

a. Determine the reduced snow load because of the roof pitch in kN/m².

b. What is the snow load per metre run on one roof truss?

c. What is the total snow load on one roof truss?

d. What is the load per metre run on the supporting wall assuming that the loads from the trusses can be considered to be uniformly distributed?

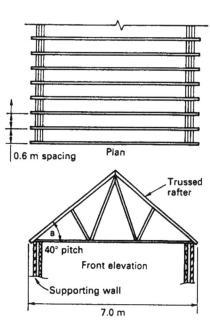

Figure 4.4a
Elevation and plan of roof structure

Solution

a.   Pitch of roof, $a = 40°$

    Load on pitched roof $= (60 - a)/30 \times 0.75$
                        $= (60 - 40)/30 \times 0.75$

Answer   **$= 0.5 \text{ kN/m}^2$**

b. Because the roof trusses are spaced at 0.6 m, the load supported per metre run by one truss is as shown in *figure 4.4b*.

Load per metre on truss $= 0.5 \times 0.6$
Answer   **$= 0.3 \text{ kN/m}$**

Comment – Note how the load acts vertically downwards (due to gravity), and that the one metre dimension is measured horizontally rather than parallel to the slope of the rafter.

c. For a 7 m wide roof:

    Total load per truss $= 0.3 \times 7.0$
Answer   **$= 2.1 \text{ kN}$**

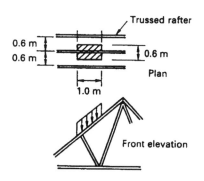

Figure 4.4b
Snow load per metre run on rafter

d. The load per metre run along the supporting wall is shown in *figure 4.4c.*

$$\text{Load on wall} = 0.5 \times 3.5$$

Answer $= \textbf{1.75 kN/m}$

Comment – For structural design purposes the above snow loads must be added to the roof dead loads. Also if limit state design is used the loads must be multiplied by the safety factors given in section 4.5.

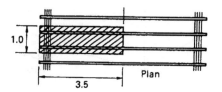

Figure 4.4c
Snow load per metre run on wall

4.4 Wind loads

Wind loads probably cause more structural failures than any other type of load. This is because wind loads act **normal** to building surfaces and can produce either positive pressures or negative suctions. In other words, they can apply loads to structures in unexpected directions which can sometimes catch the designer unawares. It means that structures must be designed to resist **horizontal** forces as well as vertical forces from gravity. In addition, lightweight structures and roofs can be subject to **uplift** forces from wind, and so they must be adequately held down.

Like snow loads, wind loads vary throughout the country and are based on meteorological data. Factors which must be taken into account include:

- Geographical location
- Degree of exposure
- Building height and size
- Building shape
- Time of exposure
- Wind direction in relation to the structure
- Positive or negative pressures within the building.

The above factors are taken into account using a range of charts and tables in the standard, however, their detailed use is beyond the scope of this book. The magnitude of the wind pressure on the side of a building is proportional to the square of the wind speed, and consequently quite wide variations are possible. The range can be from about 0.25 kN/m² for a low building in London up to 2.0 kN/m² for a tall building near Edinburgh.

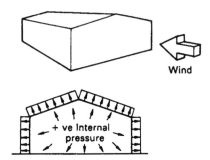

Figure 4.5a
Maximum wind uplift cast

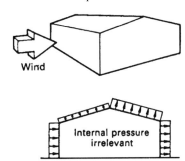

Figure 4.5b
Maximum wind sway case

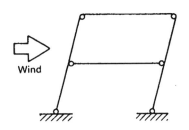

Figure 4.6
The failure of a structure designed to resist only vertical loads

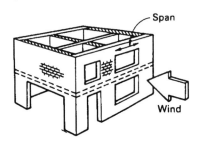

Figure 4.7a
Walls parallel to the wind direction provide stability

For structural design purposes, it may be necessary to consider several load cases produced by the wind blowing from different directions. *Figure 4.5a* shows the effect of wind blowing on the end of a factory building. This is the worst case for uplift. The structural designer must ensure that the structural components are both heavy enough, and adequately connected together, in order to resist the uplift pressure.

Figure 4.5b shows the effect of wind blowing on the side of the same building. This is the worst case for sidesway. The structure must be provided with the means to resist the sidesway forces, and some of the methods of doing this are described in the next section.

4.4.1 *Structural forms for resisting wind loads*

It is vital that the horizontal forces that result from wind loads are given careful consideration in structural design. The designer must provide a logical and coherent path to transmit the forces safely down to the ground. There is the danger that a structure, which is designed to support only vertical loads, will collapse as shown in *figure 4.6* under the influence of horizontal wind loads. The lateral stability of a structure is very important and must be considered at the very beginning of a design – not left as an afterthought. One approach is to provide a structural frame which has rigid joints. This is the approach usually adopted for the factory portal-frame buildings shown in the previous section. Other structural solutions include the building of rigid infill panels between the frame members and the provision of diagonal bracing members. *Figures 4.7a–f* show some structural forms that can resist horizontal wind forces.

Figure 4.7a shows the shell of a small domestic masonry building. The precise behaviour of such a building is complex, however, in simplistic terms the walls which face the wind span horizontally to transmit the wind force to the perpendicular walls which provide the stability. These are shown hatched. Provided there are enough bracing walls in each direction, the building will resist wind from any direction. It is rare for actual calculations to be performed for such small structures, however, the National Building Regulations specify minimum wall lengths and maximum opening sizes to ensure that the bracing walls are adequate.

It is also a requirement to provide adequate ties between the walls and the intermediate floors and the roof. This is to prevent the walls from being sucked out by negative wind pressures.

The skeleton of a portal framed factory building is shown in *figure 4.7b*. When the wind blows from the direction shown, the gable posts span vertically between the portal rafter and the base slab. The horizontal roof bracing then acts as a **wind girder** to transfer the force out to the side walls of the building. Finally the vertical bracing transmits the force down to the ground. When the wind blows from a perpendicular direction, stability is maintained by the rigid joints of the portal frames.

It is particularly important to consider the stability of tall buildings. *Figure 4.7c* shows one in which the wind forces are resisted by **shear walls** at each end of the building. They are normally made from reinforced concrete. These must be securely anchored at foundation level so that they can act as vertical cantilevers.

Figure 4.7d shows a similar concept where the shear walls have been replaced by diagonal bracing.

Stability can be obtained by anchoring the building back to one or more internal **cores**. A single central core is shown in *figure 4.7e*. Cores are normally made from reinforced concrete, and they usually act as a fireproof housing for lifts, stairs and service ducts.

Shells, **domes** and **arches** (*figure 4.7f*) are structural forms that usually have high lateral strength, and hence are good for resisting horizontal wind loads. (A shell, as the name suggests, is a thin, curved membrane structure.)

4.5 Loads from liquids and soils

Other sources of horizontal loads on structures are liquids and soils.

4.5.1 *Liquids*

Liquids produce horizontal forces in structures such as swimming pools, storage tanks and dams. We need to understand the relationship between the density of the liquid, its depth, and the horizontal forces produced. We know from simple physics that the pressure in a liquid, at any depth, is the same in all directions. It is

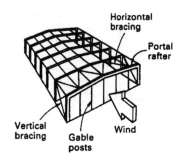

Figure 4.7b
Bracing of a portal frame building

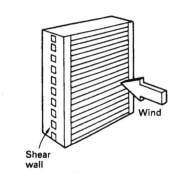

Figure 4.7c
A tall building supported by shear walls

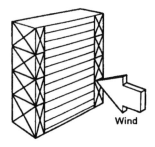

Figure 4.7d
A tall building with diagonal bracing

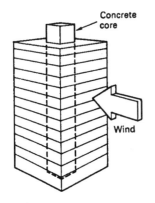

Figure 4.7e
A tall building with a central core

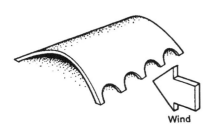

Figure 4.7f
Arched structures have inherent lateral strength

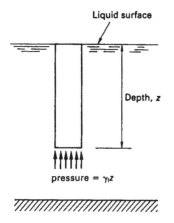

Figure 4.8
Water pressure at depth z is equal to the weight of the column of liquid above

known as **hydrostatic pressure**. Hydrostatic pressure, p, increases linearly with depth, and is proportional to the density or unit weight of the liquid. This is easy to visualise if you imagine the pressure at, say, depth z being caused by having to support the weight of the column of liquid above it. See *figure 4.8*.

Thus   hydrostatic pressure, $p = \gamma_l z$

where

$$\gamma_l = \text{unit weight of liquid}$$
$$= 9.81 \text{ kN/m}^3 \text{ for fresh water}$$
$$\text{(The unit weight is, of course,}$$
$$\text{equal to the density multiplied}$$
$$\text{by gravity, } g)$$

The hydrostatic pressure on the vertical wall of a 2.4 m deep swimming pool therefore increases with depth and produces the triangular distribution shown in *figure 4.9a*.

$$\text{Pressure at 2.4 m depth} = 9.81 \times 2.4 = 23.54 \text{ kN/m}^2$$

The **hydrostatic force** produced is simply the **average** pressure multiplied by the area over which it acts. For each metre length of wall this is as follows:

$$\text{Force on wall} = 23.54 \times 1/2 \times 2.4$$
$$= 28.25 \text{ kN/m}$$

The position of the resultant of this force occurs at the centroid of the triangular pressure distribution – i.e. a third of the depth up from the bottom. See figure 4.9b. This is used to evaluate the bending moment at the base of the wall:

$$\text{Bending moment} = \text{force} \times \text{distance}$$
$$= 28.25 \times 2.4/3$$
$$= 22.6 \text{ kNm/metre of wall}$$

4.5.2 Soils

In the case of soils, horizontal pressures result wherever a structure supports a vertical change in ground level. Such a structure is known as a **retaining wall**. *Figure 4.10* shows a cross-section through a typical retaining wall. Soils possess a property known as **shear strength**. This is the property which enables soil to be formed into a stable slope or embankment (unlike liquids!). Because of shear strength, the pressure on retaining walls is reduced from the simple hydrostatic value:

$$\text{Horizontal pressure from soil} = k_a \gamma_s z$$

where k_a = **active pressure coefficient**
γ_s = unit weight of soil (i.e. about 20 kN/m^3)
z = height of wall

The value of k_a should be determined from special shear strength tests on the soil, but in the absence of further information **a value of k_a = 1/3 is often used**. Further information on this topic can be obtained from any text book on soil mechanics. The above equation indicates that the pressure increases linearly with depth, and hence the resulting pressure distribution is also triangular (see *figure 4.10*).

It is important that retaining walls are provided with adequate drainage. If water is allowed to build up behind a retaining wall it has two effects:

- The horizontal pressure from the soil is reduced because the unit weight of the submerged soil is effectively reduced due to buoyancy.
- Hydrostatic pressure from the water must be added to the soil pressure.

The net effect of these changes is to produce a significantly higher horizontal pressure on the wall. For the case where the water level has risen to the full height behind a wall, the pressure becomes:

$$\text{Pressure from soil and water} = k_a(\gamma_s - \gamma_w)\, z + \gamma_w z$$

where γ_w = unit weight of water
= 9.81 kN/m^3

Example 4.6

A section through the wall of a concrete reservoir is shown in *figure 4.11a*. Estimate the pressure distribution on the wall, and the bending moment at the base of the wall, for the following three load cases:

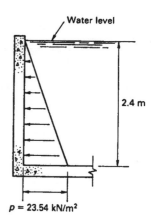

Water level

2.4 m

$p = 23.54$ kN/m^2

Figure 4.9a
Hydrostatic pressure distribution on the side-wall of a swimming pool

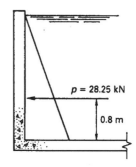

$p = 28.25$ kN

0.8 m

Figure 4.9b
The magnitude and position of the resultant hydrostatic force

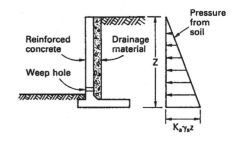

Reinforced concrete

Drainage material

Weep hole

Pressure from soil

z

$K_a \gamma_s z$

Figure 4.10
A typical retaining wall to support soil at a change in ground level

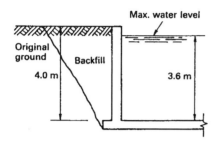

Figure 4.11a

A section through a reservoir wall

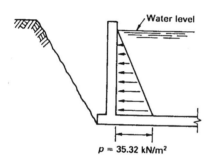

$p = 35.32$ kN/m²

Figure 4.11b

The pressure distribution from the water

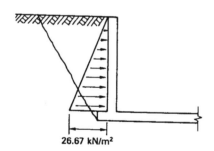

26.67 kN/m²

Figure 4.11c

The pressure distribution from the soil

a.　reservoir full of water but soil backfill not in place

b.　backfill in place but reservoir empty and good drainage

c.　as b. but no drainage so that water rises to top of soil level.

Solution

a.　Water pressure at base = $\gamma_1 z$ = 9.81 × 3.6

= 35.32 kN/m²

Answer　　**see *figure 4.11b***

Force = 35.32/2 × 3.6 = 63.57 kN

Answer　　**Moment** = 63.57 × 3.6/3 = **76.28 kNm**

b.　Assume:　　　　γ_s = 20 kN/m³

k_a = 1/3

Soil pressure at base = $k_a \gamma_s z$ = 1/3 × 20 × 4.0

= 26.67 kN/m²

Answer　　**see *figure 4.11c***

Force = 26.67/2 × 4.0 = 53.33 kN

Answer　　**Moment** = 53.33 × 4.0/3 = **71.11 kNm**

c.　Pressure from soil and water = $k_a(\gamma_s - \gamma_w) z + \gamma_w z$

= 1/3 × (20 − 9.81 × 4.0 + (9.81 × 4)

= 13.59 + 39.24 = 52.83 kN/m²

Answer　　**see *figure 4.11d***

Force = 52.83/2 × 4.0　= 105.66 kN

Answer　　**Moment** = 105.66 × 4.0/3 = **140.88 kNm**

Comment – The lack of adequate drainage has doubled the pressures on the wall.

4.6 Loads and limit state design

The dead, imposed and wind loads discussed so far are the best realistic estimates of the loads that can occur on the structure throughout its life. In limit state design these are termed **characteristic loads**:

$$G_k = \text{characteristic dead load}$$
$$Q_k = \text{characteristic imposed load}$$
$$W_k = \text{characteristic wind load}$$

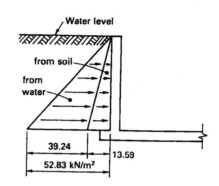

Figure 4.11d
Increased pressure due to the rise in water level behind the wall

We saw in *example 1.1* that, for checking ultimate strength, these loads must be increased by multiplying them by safety factors. These are termed **partial safety factors for loads**, γ_f The result is the **design load**.

Thus   Design load = characteristic load $\times \gamma_f$

The values of the partial safety factors for loads depend upon the type of loading and the way in which loads are combined. For design in most structural materials the values of γ_f given in *table 4.4* are used.

Table 4.4
Partial safety factors for loads, γ_f

Load combination	Dead	Imposed	Wind
Dead and imposed	1.4* or 1.0	1.6*	–
Dead and wind	1.4 or 1.0	–	1.4
Dead and imposed and wind	1.2	1.2	1.2

*Eurocodes give these values as 1.35 and 1.5 respectively.

Loads from liquids and earth pressure use the same factors as dead load.

Where two values are given for dead load then the most adverse value must be used. Also the imposed load is only to be present if it makes the loading condition more adverse. Consider the forest fire observation tower shown in *figure 4.12* with wind blowing on the side.

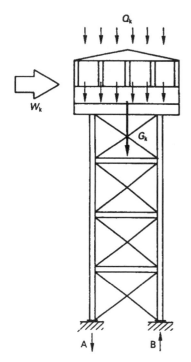

Figure 4.12
A forest fire observation tower with wind forces on the side

To obtain the **maximum compressive** design load in the support leg at B, two load combinations should be checked and the larger value used in subsequent design calculations:

$$1.4G_k + 1.6Q_k$$

or

$$1.2G_k + 1.2Q_k + 1.2W_k$$

To obtain the **maximum tensile** design load in the support leg at A, we must minimise the effect of dead and imposed load by using the following combination:

$$1.0G_k + 1.4W_k$$

In other words the tower is much more likely to blow over when the tank is empty.

For large and complex structures the theoretical number of possible load combinations can be formidable. However, the experienced designer quickly learns which ones are the most critical.

When checking structures against **serviceability** limit states such as deflection or cracking, the loads are usually used unfactored.

4.7 Calculating loads on beams

Before a beam can be designed it is clearly necessary to establish the design load that it is to carry. Very often, loads are assumed to be either pure **point loads** or **uniformly distributed loads** (**UDLs**), or a combination of the two. A UDL is a load which is evenly spread out over a particular length of the beam. In practice they may not be pure point loads or pure uniformly distributed loads, but this is usually a good enough assumption for design purposes. *Figure 4.13* shows how point loads and uniformly distributed loads are represented. Care must be taken with the units of UDLs. They are usually given as a load-per-unit-length of beam – kN/m, but sometimes they may be given in the form of the total load – kN, which is then assumed to be evenly spread throughout a length of beam. Always look at the units carefully to make sure that you are interpreting the load correctly.

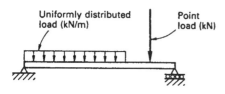

Figure 4.13
The representation of uniformly distributed loads and point loads on a beam

4.7.1 Loads from one-way spanning floors

Certain types of floor and roof slab construction have a clear direction of span. Examples of this type include timber joists and precast concrete beam-and-pot construction. See *figure 4.14*. It is easy to see that all loads (dead and imposed) are carried by the supporting beams which run perpendicular to the direction of span. The beams which run parallel to the direction of span, in this case, carry no load other than their own self-weight. Although the load is applied to the supporting beams in the form of a series of point loads (one from each joist in the case of the timber floor), they can be considered close enough together to form a uniformly distributed load.

The load on each beam is clearly related to the area of floor or roof that it supports. The following example shows how beam loads are calculated.

Example 4.7

Figure 4.15a shows a plan view of a corner of a typical floor in a multi-storey building. The arrows indicate the direction of span of the floor. Determine the design loading and beam reactions for all the beams numbered *B1* to *B5* for the case of dead + imposed load.

Use of building – general office
Floor construction – as **example 4.2** – i.e. 4.11 kN/m^2
Perimeter wall construction – as **example 4.3** – i.e. 4.77 kN/m
Estimated self-weight of beams – 0.6 kN/m

Solution

Design loads

The above loads are characteristic loads. *Table 4.4* indicates that we must multiply the characteristic dead load, G_k, by 1.4 and the characteristic imposed load, Q_k, by 1.6 in order to obtain design loads.

Design dead load (floor) = 1.4 × 4.11 = 5.75kN/m^2
Design dead load (wall) = 1.4 × 4.77 = 6.68 kN/m^2
Design dead load (beams) = 1.4 × 0.6 = 0.84 kN/m

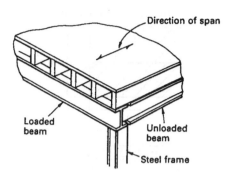

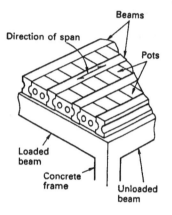

Figure 4.14
Examples of one-way spanning floors

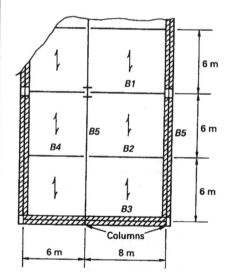

Figure 4.15a
The plan view of the corner of a multi-storey building floor

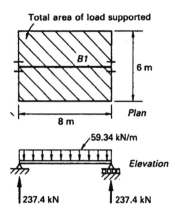

Total area of load supported

B1

6 m

Plan

8 m

59.34 kN/m

Elevation

237.4 kN   237.4 kN

Figure 4.15b
Loading on beams *B1* and *B2*

From *table 4.3* for general office loading

$$\text{Design imposed load (floor)} = 1.6 \times 2.5 \quad = 4.00 \text{ kN/m}^2$$

$$\text{Total design floor load} = 5.75 + 4.00 = 9.75 \text{ kN/m}^2$$

Beam B1

The beam supports a total width of floor of 6.0 m – see *figure 4.15b*. For each metre length of beam the load supported is as follows:

$$\text{Load from floor} = 9.75 \times 6.0 = 58.50 \text{ kN/m}$$
$$\text{Self-weight of beam} \qquad = \underline{\ 0.84 \text{ kN/m}}$$

Answer   Total design UDL = 59.34 kN/m

From symmetry, the reactions at each end of all the beams must be the same, and equal to half the total load on the beam.

Answer   Reaction $= \dfrac{(59.34 \times 8)}{2} = $ **237.4 kN**

See *figure 4.15b*.

Beam B2

Answer   As Beam *B1*

Beam B3

The beam supports a width of floor of 3 m plus the weight of the perimeter wall.

$$\text{Load from floor} = 9.75 \times 3.0 = 29.25 \text{ kN/m}$$
$$\text{load from wall} \qquad = 6.68 \text{ kN/m}$$
$$\text{self-weight of beam} \qquad = \underline{\ 0.86 \text{ kN/m}}$$

Answer   Total design UDL = 36.79 kN/m

Answer   Reaction $= \dfrac{36.90 \times 8}{2} = $ **147.6 kN**

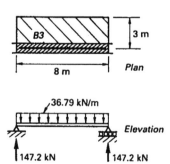

B3

3 m

Plan

8 m

36.79 kN/m

Elevation

147.2 kN   147.2 kN

Figure 4.15c
Loading on beam *B3*

See *figure 4.15c*

Beam B4

Answer   Total design UDL per metre is the same as beams B1 and B2 – i.e. 59.34 kN/m

Answer   Reaction $= \dfrac{59.34 \times 6}{2} = \textbf{178.0 kN}$

See *figure 4.15d*

Beam B5

The beam supports the weight of the perimeter wall. No floor load is carried directly, but the reaction from beam *B2* acts as a point load at mid-span.

$$
\begin{aligned}
\text{load from wall} &= 6.68 \text{ kN/m} \\
\text{self-weight of beam} &= \underline{0.86 \text{ kN/m}}
\end{aligned}
$$

Answer   Design UDL = 7.54 kN/m

Answer   Design point load = 237.4 kN

Answer   Reaction $= \dfrac{(7.54 \times 12) + 237.4}{2} = \textbf{163.9 kN}$

See *figure 4.15e*

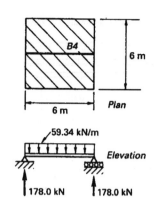

Figure 4.15d
Loading on beam *B4*

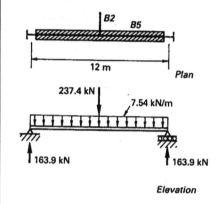

Figure 4.15e
Loading on beam *B5*

Example 4.8

Figure 4.16a shows a typical private double garage. Access is to be provided to the roof so that it can be used as a patio. Determine the loading on the two steel beams – *B1* and *B2*. (*In a later example we will actually determine the size of these two beams.*)

Solution

Design loads

The imposed loading on the roof must be taken as the worst of either snow loading or 1.5 kN/m² (because access is provided). We know from *section 4.3.2* that the snow loading is only of the order of 0.75 kN/m², and hence will not be critical.

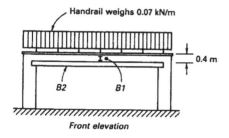

Handrail weighs 0.07 kN/m

0.4 m

B2　　B1

Front elevation

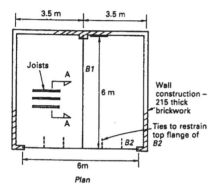

3.5 m　　3.5 m

Joists

A

B1

6 m

A

A

Wall
construction –
215 thick
brickwork

Ties to restrain
top flange of
B2

B2

6m

Plan

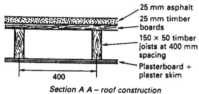

25 mm asphalt
25 mm timber
boards
150 × 50 timber
joists at 400 mm
spacing
Plasterboard +
plaster skim

400

Section A A – roof construction

Figure 4.16a

Elevation, plan and roof detail of double garage

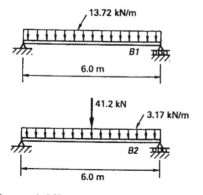

13.72 kN/m

B1

6.0 m

41.2 kN

3.17 kN/m

B2

6.0 m

Figure 4.16b

Loading on beans *B1* and *B2*

The first step is to calculate the characteristic dead load of one square metre of roof. Refer to *tables 4.1* and *4.2* for weights of the various materials.

asphalt	– 0.45 × 25/19	= 0.59
boards	– 6.0 × 0.025	= 0.15
joists	– 6.0 × 0.15 × 0.05 × 1/0.4	= 0.11
plasterboard + skim		= 0.15
characteristic roof dead load		= 1.00 kN/m^2

Comment – The joist loading requires some explanation. The 6.0 is clearly the unit weight of timber. This is then multiplied by the volume of timber in 1 square metre of roof. 0.15 × 0.05 × 1 is the volume of one joist. 1/0.4 is the number of joists in one square metre – based on a spacing of 400 mm.

Design dead load $= 1.4 \times 1.00 = 1.4 \text{ kN/m}^2$
Design imposed load $= 1.6 \times 1.5 = 2.4 \text{ kN/m}^2$
Total roof design load $= 3.8 \text{ kN/m}^2$

Self-weight (beam) $= 1.4 \times 0.3 = 0.42 \text{ kN/m}$
Dead load of walls $= 1.4 \times 22.0 \times 0.215 = 6.62 \text{ kN/m}^2$
Self-weight (handrail) $= 1.4 \times 0.07 = 0.10 \text{ kN/m}$

Beam B1

Comment – The load from the joists actually forms a series of point loads along beam B1 – one every 400 mm. However, this is an example where the load is assumed to be a uniformly distributed load (UDL).

Answer　　**UDL** $= (3.8 \times 3.5) + 0.42$
$= \textbf{13.72 kN/m}$

Beam B2

This supports a UDL from the 0.4 m high strip of brickwork and the handrail plus a point load from beam *B1*.

Answer　　**Design point load** $= 13.72 \times 6/2 = \textbf{41.2 kN}$

UDL from brickwork $= 6.62 \times 0.4 = 2.65$
Handrail $= 0.10$
Self-weight $= 0.42$

Answer Design UDL $= \textbf{3.17 kN/m}$

See *figure 4.16b*

4.8 Summary of key points from chapter 4

1. The three principal types of load to be considered are **dead**, **imposed** and **wind**.
2. Dead loads arise from the weight of permanent parts of a structure such as the weight of the structural members. For linear members such as beams this is usually calculated in kN/m, and for flat members such as walls and floors it is usually calculated in kN/m^2.
3. Imposed (or live) loads are variable and result from the 'users' of a structure such as people, cars or library books. Characteristic values are given in Standards in kN/m^2.
4. Wind loads are important because they can impose horizontal forces on a structure, and some way of resisting these forces must be provided.
5. Liquids and soils can also impose horizontal forces.

$$\text{hydrostatic liquid pressure} \quad p = \gamma_l \times \text{depth}$$
$$\text{horizontal soil pressure} \quad = k_a \times \gamma_s \times \text{depth}$$

It is important to provide drainage at the back of retaining walls to prevent the build-up of water pressure.
6. Each load must be multiplied by a **partial safety factor for loads**, γ_f, before it becomes a **design load** for use in limit state design.
7. Loads on beams are often either **point** loads (kN) or **uniformly distributed** loads (kN/m).

4.9 Exercises

E4.1 A factory workshop floor is made from steel plates 6 mm thick.
a) Determine the characteristic dead load of the floor in kN/m^2.
b) What is the total design load for the floor if imposed load is included?
a) *characteristic dead load = 0.42 kN/m^2*
b) *design load = 8.59 kN/m^2*

E4.2 A large tank in a chemical plant has vertical walls 4.5 metres high. It is full to the top with a liquid which has a unit weight of 8.5 kN/m^3.

a) Sketch the shape of the pressure distribution on the wall and indicate the peak value.
b) What is the total characteristic horizontal force on the wall for each metre length of wall.
c) What bending moment does this force exert about the base of the wall?

a) peak pressure $= 38.25 \; kN/m^2$
b) force/m $= 86.1 \; kN/m$
c) bending moment $= 129.2 \; kNm/m$

E4.3 Each tower of a two-tower skyscraper is about 60 metres square and has 110 storeys. Each storey has a height of about 4 metres. It can be assumed that the floors are made from concrete 200 mm thick and the walls weigh an average of 1.0 kN/m². It is used for general offices. Give a rough estimate of the total design load that must be carried by each tower's foundations if all the floors are fully loaded. (Note: In practice the design load can be reduced somewhat because it is assumed that not all floors will be fully loaded at the same time. Take care: 60 metres square is not the same as 60 square metres!)

Approx. $8.4 \times 10^6 \; kN$

E4.4 *Figure 4.17* shows a cross-section through part of the roof of a building which is used for car parking. Calculate the design load on the steel beam in kN/m.

design load on beam $= 743 \; kN/m$

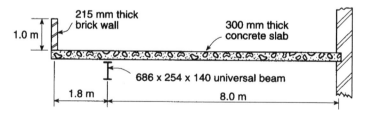

Figure 4.17
Cross-section through car park slab

Chapter 5

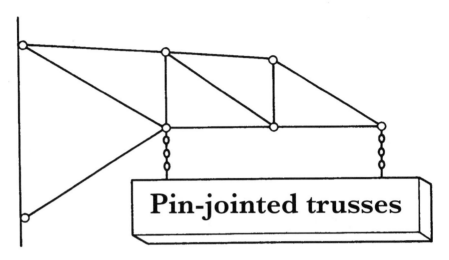

Pin-jointed trusses

Topics covered

Examples of pin-jointed frame structures
Analysis of frames to obtain the compressive or tensile
*   force in each member*
Identification of frames which are unstable or contain
*   redundant members*
Unloaded members
Space frames

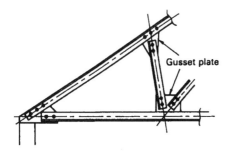

Figure 5.1
Typical bolted 'gusset plate' connections –
assumed to be pinned for the purpose of
analysis

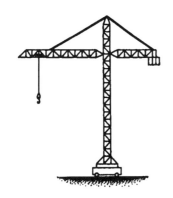

Figure 5.2a
A tower crane

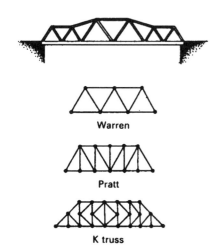

Figure 5.2b
Tubular steel footbridge and truss types

5.1 Introduction

A **pin-jointed truss** is a structure made up from separate components by connecting them together at pinned joints or **nodes**, usually to form a series of triangles. In practice, modern structures rarely have true pinned joints (*figure 5.1*); however, if pins are assumed it simplifies the analysis and results in reasonably accurate and practical solutions in most cases. It is only with very large or repetitive structures that a more complex computer analysis is really justified. Initially we shall be restricting the analysis to **plane** trusses, which means that all of the members and loads lie in one flat plane.

This chapter is largely concerned with the **analysis** of trusses, which in this context means determining the magnitude of the tensile and compressive forces in the truss members. Tensile members are called **ties** and compression members are called **struts**. *Chapters 6* and *8* continue the design process by looking at how we determine the dimensions of suitable members to carry these forces, and *chapter 11* looks at how we might connect the members together. *Chapter 14* shows how to calculate the deflection of trusses.

In general, trusses are good for carrying relatively light loads over large distances, and they tend to be relatively rigid, so that such structures do not sway or deflect as much as solid structures of equivalent weight. They do, however, take up more room than solid structures of the same strength. They are an efficient way of minimising material weight, but their manufacture can be labour intensive and hence expensive. The advent of more automation and robotics in the manufacturing process is likely to give pinpointed structures an economic boost.

5.2 Examples of trusses

Many types of crane (*figure 5.2a*) use trusses because of the combination of lightness and rigidity that they provide. The crane jibs are, in fact, made from three or four plane trusses joined at the corners to form a 'box' structure, which provides strength in all directions.

The footbridge (*figure 5.2b*) is fabricated from steel tubing with welded joints, which gives a clean pleasing appearance. The most economic depth for a truss is usually around one-sixth to one-eighth of the span.

Several styles of truss have evolved and some of these are shown. Under certain circumstances a particular style can be the most economic. The usual aim is to minimise the length of the compression members because, as we shall see later, the strength of a strut is inversely proportional to the square of its length, but with a tension member the length is irrelevant to its load-carrying capacity.

The transmission tower (*figure 5.2c*) is, like the crane jibs, made up from plane trusses. The principal force on tower structures is often horizontal wind loading, which, when blowing on one face, would be divided between the two perpendicular trusses. Note how the pattern of bracing changes as the tower narrows towards the top. This is done to restrict the length of the struts and at the same time to achieve a well-proportioned set of triangles. Angles between adjacent members should be kept between 30° and 60° where possible.

Although the tower looks a relatively simple structure, the very high numbers of identical structures justify a detailed and complex analysis. They are even subjected to full-scale destructive testing in order to fully prove the design. They must be designed to cover such incidents as the cables being weighed down by ice and snow, and also a cable snapping on one side of the tower, which subjects the tower to a large twisting torque.

Timber roof trusses (*figure 5.2d*) are now regarded as the most economic way of providing pitched roofs of spans up to say 12 m. They have largely displaced the more traditional use of heavy timber beams or **purlins** supporting separate sloping rafters in routine house building. The trusses are closely spaced at 600 mm centres and each one is light enough to be manhandled into position.

The simple farm gate (*figure 5.2e*) is a truss structure. The only load is its own self-weight, and the top and bottom rails together with the internal vertical and diagonal members actually form the structure (the other rails are to keep the sheep in the field!).

Figure 5.2c
A transmission tower

Figure 5.2d
A timber trussed rafter

Figure 5.2e
A field gate

5.3 Forces in members

Although not always possible, it is best to apply loads to a truss only at the node points. If this is the case, and frictionless pinned

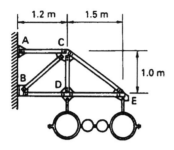

Figure 5.3a

A cantilever bracket for supporting pipes

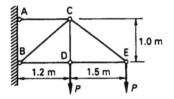

Figure 5.3b

The structural model

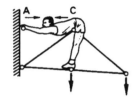

Figure 5.3c

Member AC

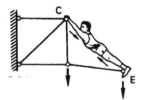

Figure 5.3d

Member CE

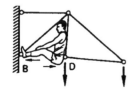

Figure 5.3e

Member BD

joints are assumed, then the members will be subject only to axial compression or tension. There is **no** bending or shear on the individual members. The load from the actual weight of the truss members presents a problem here, since their weight is spread throughout the structure and not just concentrated at the nodes. The usual practice is to use increased values for the nodal loads during the calculations to compensate for the self-weight of the truss. The size of the increase required is initially a guess based on experience, but should be checked at the end of the design when the actual weights of the members are known. Occasionally a redesign is required if the original guess turns out to be an underestimate.

5.3.1   *Finding compression or tension by inspection*

In most cases it is easy to determine whether a truss member is in compression or tension simply by looking at the structure. This is an important skill that should be developed as it can expose embarrassing fundamental errors in numerical calculations.

The key is to imagine each member in turn to be cut or removed from the structure while under load. This will produce an unstable structure or **mechanism**. If the nodes at each end of the member would move closer together (and hence have to be pushed apart to prevent collapse) then the member is in compression. This is shown on a diagram by placing a small arrow close to the end of the member as follows: ⟵——⟶, i.e. pushing the nodes apart. Conversely, if the nodes at each end of the member would move apart (and hence have to be pulled together to prevent collapse) then the member is in tension. This is indicated as: ⟶——⟵, i.e. pulling the nodes together. This is best illustrated by considering an example.

Example 5.1

Figure 5.3a shows one of a series of cantilever brackets used to support pipes in a chemical plant. Put arrows on a diagram of the structure to indicate whether the force in each member is tensile ⟶⟵ or compressive ⟵⟶.

Solution

The first step is to transform the diagram of the actual structure into a structural model with loads (*figure 5.3b*).

The procedure is to cut mentally each member in turn and to picture how the structure would deform.

The perimeter members are easy.

Member AC – C tries to move away from A so member AC must pull – hence tension (*figure 5.3c*).

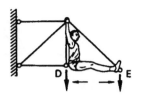

Figure 5.3f
Member DE

Member CE – Also tension (*figure 5.3d*).

Member BD – D tries to move towards B so BD must push – hence compression (*figure 5.3e*).

Member DE – Also compression (*figure 5.3f*).

The internal members can be a little more difficult.

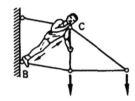

Figure 5.g
Member BC

Member BC – With BC removed the rectangle ACDB wants to form into a parallelogram with C moving closer to B. BC must therefore push to prevent this – hence compression (*figure 5.3g*).

Member CD – D tries to drop away from C so CD must pull – hence tension (*figure 5.3h*).

Answer   see *figure 5.3i*.

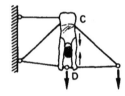

Figure 5.3h
Member CD

For structures with more complex loads it is not always possible to establish whether a member is in tension or compression simply by inspection, and the numerical values of the forces must be found.

5.3.2 Method of joints

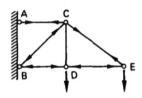

Figure 5.3i
The representation of tension and compression

We saw in *chapter 2* that the external loads and reactions on a structure must be in equilibrium. However, for each individual joint, the vector sum of the member forces plus any applied loads must also be in equilibrium.

Point A in *figure 5.4a* symbolises a pinned joint in a truss. If this system is in equilibrium, i.e. not moving, then the various forces meeting at A must cancel each other out. The force triangles (*figure 5.4b*) show the unknown forces F_{AB} and F_{AC} split into their horizontal and vertical components. The directions of the forces have been deduced by inspection – i.e. F_{AB} pulling away from A and F_{AC} pushing towards A. However, if they had been deduced

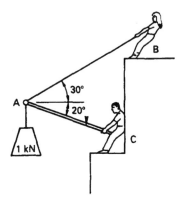

Figure 5.4a
Consider the equilibrium of joint A

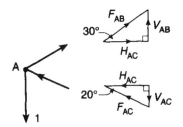

Figure 5.4b
Forces resolved into vertical and horizontal components

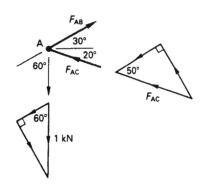

Figure 5.4c
Forces resolved perpendicular to member AB

wrongly it would simply result in negative values emerging from the calculations. It is important that we can use the direction of the vector arrows to determine whether the member is in tension or compression. Imagine the arrow at the other end of the member to be in the opposite direction. It should then be clear whether the member is in compression ⟵——⟶, or tension ⟶——⟵

Thus for vertical equilibrium:

$$(\Sigma V = 0) \qquad 1 \text{ kN} = V_{AB} + V_{AC}$$

and for horizontal equilibrium:

$$(\Sigma H = 0) \qquad H_{AB} = H_{AC}$$

Also
$$V_{AB} = F_{AB} \times \sin 30°$$
$$H_{AB} = F_{AB} \times \cos 30°$$
$$V_{AC} = F_{AC} \times \sin 20°$$
$$H_{AC} = F_{AC} \times \cos 20°$$

The equations can be written:

$$F_{AB} \sin 30° + F_{AC} \sin 20° = 1 \qquad [5.1]$$

$$F_{AB} \cos 30° - F_{AC} \cos 20° = 0 \qquad [5.2]$$

From these two simultaneous equations the two unknown forces can be found.

Giving:
$$F_{AB} = 1.23 \text{ kN tension}$$
$$F_{AC} = 1.13 \text{ kN compression}$$

The need to solve simultaneous equations can usually be avoided, however, if the forces are resolved perpendicular to one of the unknown forces, rather than vertically and horizontally. This has the effect of eliminating that force from the equations, because a force does not have a component perpendicular to itself.

For example resolve forces perpendicular to AB:

From *figure 5.4c*

$$1 \times \sin 60° = F_{AC} \sin 50° \qquad \text{hence } F_{AC}$$

and now resolve parallel to AB, F_{AC} now being a known quantity:

$$F_{AB} = 1 \times \cos 60° + F_{AC} \cos 50° \qquad \text{hence } F_{AB}$$

The same problem can be solved even more easily and quickly by a scale drawing of the vectors of the forces meeting at A. The length of each line is drawn proportional to the force, so a suitable scale must be determined. The vectors are drawn parallel to each force, starting with the known value. The magnitudes of the two unknown forces are determined by intersection. Starting from each end of the known force vector, draw lines parallel to the two unknown forces until they intersect (*figure 5.4d*). The fact that the vectors form a closed circuit ensures that equilibrium has been satisfied. Obviously the horizontal components of F_{AB} and F_{AC} must cancel out, and likewise with the sum of the vertical components of all the forces.

You can see that the arrows form a continuous loop around the circuit. This means that if you are uncertain about the directions of the unknown forces, they can be added by following the direction of the known force. They can then be transferred to the diagram of the structure in the form of *figure 5.3i*. Remember, however, that the arrows at each end of a particular member are in opposite directions. In this case the arrows are added to the appropriate members adjacent to joint A.

More than three forces at one point can be dealt with, but not more than two of them can be unknown. This is because there are only two independent equilibrium equations available. This is illustrated in the following example.

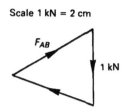

Scale 1 kN = 2 cm

F_{AB}

1 kN

Figure 5.4d
Force triangle at A

Example 5.2

If, in *example 5.1*, the brackets occur every 5 m, and the total design load of the pipes is 3.8 kN/m, determine the value of the forces in all the members:

a. by calculation
b. by scale drawing.

Ignore the self-weight of the truss.

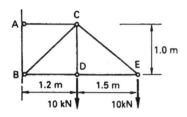

Figure 5.5a
Length of pipe supported by one bracket

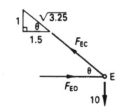

Figure 5.5b
The structural model

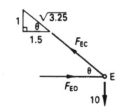

Figure 5.5c
The equilibrium of joint E

Solution

Loads

It is first necessary to determine the design load on one bracket (*figure 5.5a*).

Weight of pipes supported by each bracket = 5 × 3.8 = 19 kN

Because of symmetry of pipes, divide this weight equally between nodes D and E (*figure 5.5b*).

$$\text{load per joint} = \frac{19}{2} \; 9.5 \text{ kN}$$

Allowing for self-weight of truss, load per joint, say = 10 kN

The structural model is therefore as shown in *figure 5.5b*.

a. By calculation

Joint E

The two unknown forces, F_{EC} and F_{ED} (*figure 5.5c*), must be in equilibrium with the applied load of 10 kN.
    In other words the vertical components of all the forces must sum to zero and likewise with horizontal components.

i.e.
$$\Sigma V = 0$$
$$\Sigma H = 0$$

$(\Sigma V = 0)$ $F_{EC} \times \sin \theta = 10$

$$\frac{F_{EC} \times 1}{\sqrt{3.25}} = 10$$

Answer $\mathbf{F_{EC} = 18.0 \ kN \ tension}$

$(\Sigma H = 0)$ $F_{ED} = F_{CE} \cos \theta$

$$= \frac{18.0 \times 1.5}{\sqrt{3.25}}$$

Answer $\mathbf{F_{ED} = 15.0 \ kN \ compression}$

Comments:

1. *Because only two equations of equilibrium are available ($\Sigma V = 0$ and $\Sigma H = 0$) it is only possible to deal with two unknown forces at any joint. This is why joint E was considered first.*
2. *The initial directions of F_{EC} and F_{ED}, as indicated by the arrows on figure 5.5c were determined by inspection. If they had been wrongly chosen it would merely result in the calculation producing negative values for the forces.*
3. *With many problems it is easier to work with distances rather than angles. In the above example it was not necessary to evaluate the magnitude of the angle θ.*
4. *Vertical equilibrium was deliberately considered first because, as member ED is horizontal, it does not have a vertical component and hence is eliminated from the calculation, leaving only one unknown.*

Figure 5.5d
Joint D

Joint D

Having found the forces in all the members meeting at E, it is possible to move to joint D (*figure 5.5d*) as the force in member DE is now a known quantity. (Joint C still has more than two unknowns and so must come later.)

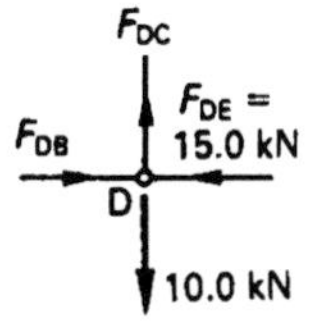

$(\Sigma V = 0)$ $F_{DC} = 10.0$ **kN**
$(\Sigma H = 0)$ $F_{DB} = 15.0$ **kN**

Joint C – See *figure 5.5e*.

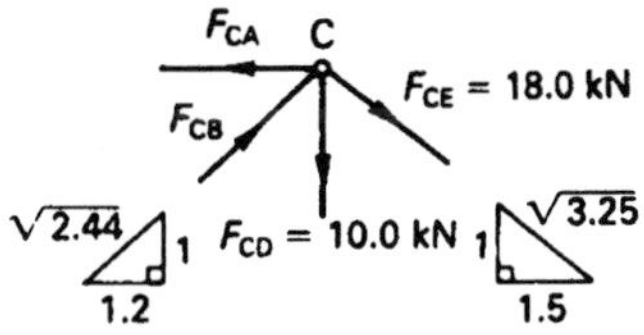

Figure 5.5e
Joint C

$(\Sigma V = 0)$ $\dfrac{F_{CB} \times 1}{\sqrt{2.44}} = F_{CD} + \dfrac{F_{CE} \times 1}{\sqrt{3.25}}$

$$= 10.0 + \frac{18 \times 1}{\sqrt{3.25}}$$

$F_{CB} = 31.2$ **kN compression**

$(\Sigma H = 0)$ $F_{CA} = \dfrac{F_{CB} \times 1.2}{\sqrt{2.44}} + \dfrac{F_{CE} \times 1.5}{\sqrt{3.25}}$

$$= \frac{31.2 \times 1.2}{\sqrt{2.44}} + \frac{18 \times 1.5}{\sqrt{3.25}}$$

$F_{CA} = 38.9$ **kN tension**

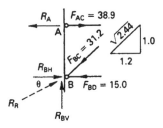

Figure 5.5f
Wall reactions

The values of all member forces have now been found. However, a designer would also need to consider the connections to the wall at A and B. Hence values for the reactions are required.

Because only a single member is connected to A, the reaction at A must be in-line with the member and of equal and opposite value. See *figure 5.5f.*

Hence $R_A = 38.9$ kN

Consider the reaction at B divided into its vertical and horizontal components.

$$R_{BV} = \frac{31.2 \times 1}{\sqrt{2.44}}$$

$$\boldsymbol{R_{BV} = 20.0 \text{ kN}}$$

$$R_{BH} = \frac{15.0 + 31.2 \times 1.2}{\sqrt{2.44}}$$

$$\boldsymbol{R_{BH} = 39.0 \text{ kN}}$$

These two components may now be converted to the resultant reaction force and its angle of action can be calculated.

$$\text{Resultant } R_R = \sqrt{(20^2 + 39^2)}$$
$$R_R = 43.8 \text{ kN}$$

$$\text{Direction } \theta = \tan^{-1} \frac{20}{39}$$

$$\theta = 27.1°$$

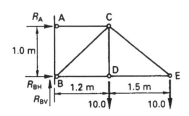

Figure 5.5g
Overall equilibrium check

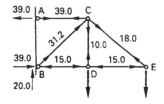

Figure 5.5h
The forces in the members

Overall equilibrium check

It is very easy to make numerical errors in calculating the forces in large trusses, but there is always enough information to make an independent check. In this case it is possible to check the reaction forces by looking at the equilibrium of the whole structure (*figure 5.5g*).

(ΣM about B)
$$(10 \times 1.2) + (10 \times 2.7) = R_A \times 1$$
$$R_A = 39.0 \text{ kN check}$$
(The previous value of 38.9 contained a small rounding error.)

$(\Sigma V = 0)$ $R_{BV} = 10 + 10 = 20.0$ kN ... check
$(\Sigma H = 0)$ $R_{BH} = R_A$ $= 39.0$ kN ... check

Answer   see *figure 5.5h.*

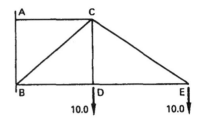

Figure 5.6a
The structural model

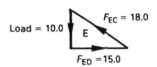

Figure 5.6b
Joint E

b. By scale drawing of a vector diagram

Careful scale drawing provides sufficient accuracy for structural design, and the likelihood of mathematical slips is reduced. A proper drawing board should be used, and the starting point is an accurate scale drawing of the structure (*figure 5.6a*).

The joints must be considered in the same order as in the previous approach – i.e. not more than two unknowns at the joint under consideration. The procedure is to draw a force triangle or polygon for each joint. A suitable scale must be determined – in this case 10 kN = 1 cm.

Joint E (figure 5.6b)

Joint D (figure 5.6c)

Joint C (figure 5.6d)

Joint B (figure 5.6e)

Comments:

1. Other than leaving the unknown forces until last, the order in which the vectors are drawn is immaterial.

2. As with the calculation approach, it is desirable to carry out an overall equilibrium check by taking moments to obtain the reactions, and comparing with the values obtained from the force triangle at joint B.

Answer   by scaling from the force polygons and noting the directions of the forces we obtain the results shown in *figure 5.5h.*

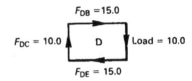

Figure 5.6c
Joint D

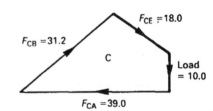

Figure 5.6d
Joint C

5.3.3 *Composite force diagram*

In the previous graphical method of joints the vector for each member needed to be drawn twice, because it was included in the force polygon at both end joints. With a **composite force**

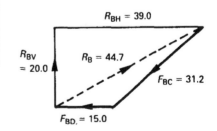

Figure 5.6e
Reactions

diagram (sometimes called a **Maxwell diagram**) this is not necessary as all the joint resolutions are combined into one diagram. To prevent confusion it is necessary to adhere to a convention known as **Bow's notation**. This is explained in the following example.

Example 5.3

Repeat the analysis of the truss shown in *example 5.2* using a composite force diagram.

Solution

Step 1 – Number all the **spaces between** the members and the external forces and reactions (*figure 5.7a*). Each force can then be identified by the two numbers on each side of it.

Comments:

1. Although the direction of the reaction at B is not yet known, it can be added approximately at this stage.

2. Number 1 covers the whole space from R_A to R_B as far as the member BC.

3. Number 2 covers the whole space from R_A to the load at E.

4. Force 1–2 therefore represents both R_A and the force in member AC.

Step 2 – Again starting at a joint with only two unknowns (joint E in this case) draw the force polygon to a suitable scale. Add the numbers to define the force by working consistently clockwise around the joint (*figure 5.7b*).

Comments:

1. The known force of 10 kN is drawn first. Going clockwise around joint E in figure 5.6b, force 2–3 is 10 kN ↓, therefore point 3 is shown **below** point 2 on the force diagram.

2. Force 3–6 represents the force in member DE and it can be seen that on the force diagram point 6 is to the right of point 3. Therefore the direction of the force in DE is →. This arrow could therefore be added to the diagram of the structure close to E, as in figure 5.5h, thus indicating that member DE is in compression.

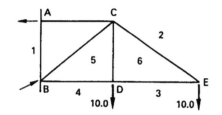

Figure 5.7a
The numbering for Bow's notation

Figure 5.7b
The first step in drawing the composite force diagram at joint E

Step 3 – Moving on to the next joint with only two unknowns (joint D) the forces are now added to the same diagram (*figure 5.7c*).

Step 4 – Joint C (*figure 5.7d*).

Step 5 – From *figure 5.7d* it can be seen that force 4–1 represents the reaction at B. Its magnitude and direction can therefore be determined.

Step 6 – Again the reactions should be checked by taking moments for the whole structure as in *example 5.2*.

Answer The magnitudes of the forces are scaled off the diagram and the directions of the forces determined either by inspection or by the method described in *step 2*. The results are as shown in *figure 5.5e*.

Example 5.4

Figure 5.8a shows a footbridge spanning 12 m across a river. All the members, except those in the lateral bracing, are 3 m long and the walkway is 2 m wide. An estimated weight for the bridge itself is 35 kN and the required characteristic imposed load is 4 kN/m². Partial safety factors of 1.4 on dead load and 1.6 on imposed load should be used.

Find the forces in all the members of a main truss.

Solution

Figure 5.8b shows the structural model of one truss. The first step is to determine the value of the loads, *P*, acting at each node. They are made up of a dead load component and an imposed load component.

Loads

Dead load – assume uniformly distributed throughout the bridge.

$$\text{Weight of bridge/m} = \frac{35}{12} = 2.92 \text{ kN/m}$$

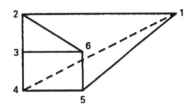

Figure 5.7c
Adding joint D

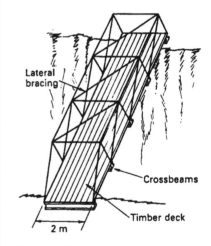

Figure 5.7d
Adding joint C

Figure 5.8a
Footbridge structure

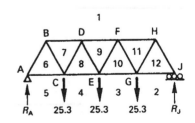

Figure 5.8b
The structural model

Nodal loads occur every 3 m, so for 3 m:

$$\text{Weight} = 2.92 \times 3 = 8.76 \text{ kN}$$

This must now be divided evenly between the two trusses:

$$\text{Dead load per truss} = \frac{8.76}{2} = 4.38 \text{ kN}$$

Imposed load – Each cross beam supports a 3 m length of deck which is 2 m wide:

$$\text{Nodal imposed load per truss} = \frac{(4 \times 2 \times 3)}{2} = 12.0 \text{ kN}$$

Now applying the safety factors

$$\text{Design load, } P = (4.38 \times 1.4) + (12.0 \times 1.6)$$
$$= 25.3 \text{ kN}$$

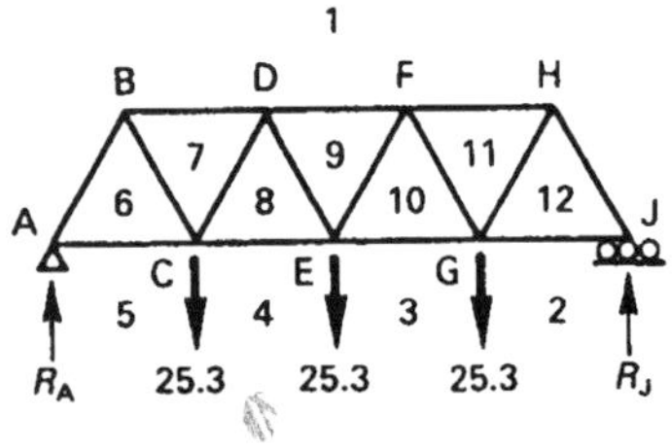

Comment – There are also loads of P/2 occurring at the end nodes A and J. These need to be included when designing the bearings and foundations, but are irrelevant to the design of the truss as they pass straight into the supports without transmitting any load to the truss. Hence they can be ignored here.

Figure 5.8c
The structural loads

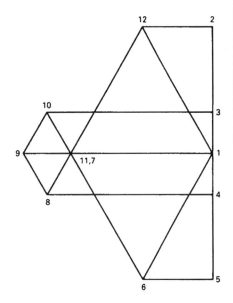

Figure 5.8d
The composite force diagram

Forces

From *figure 5.8c* it can be seen that there is no joint with only two unknown forces. In cases like this it is therefore necessary to find one of the reactions by taking moments.

$(\Sigma M$ about A)
$$(25.3 \times 3) + (25.3 \times 6) + (25.3 \times 9) = R_{\text{J}} \times 12$$

$$\boldsymbol{R_{\text{j}} = 37.95 \text{ kN}}$$

Because, in this case, both structure and loads are symmetrical both reactions should simply be half the total load.

check $$R_{\text{J}} = \frac{(3 \times 25.3)}{2} = 37.95 \text{ kN}$$

A composite force diagram can now be drawn, starting at node J and moving on to nodes H, G, F, E, D, C, B and A (*figure 5.8d*).

Comments:

1. It can be seen in this case that, as expected, the diagram is symmetrical and hence it was really only necessary to draw the half of it up to node E.

2. The last force to be drawn, 6–1, should be both parallel to member AB and also should close at point 1. This provides a check on the accuracy of drawing.

The forces, and their directions, can now be extracted from the diagram and are shown in the table below:

Answer

Member	Force (kN)	
	tension	compression
AB		43.8
AC	21.9	
BC	43.8	
BD		43.8
CD		14.6
CE		51.1
DE	14.6	
DF		58.4

5.3.4 *Method of sections*

This method is particularly suitable when it is necessary to find the force in only a few members of a truss.

In *chapter 2* we saw how the external loads and reactions acting on a structure must be in equilibrium. The **method of joints** showed that the forces acting on each joint must be in equilibrium. The **method of sections** is based on the fact that the internal forces and external loads and reactions on any part or section of a structure must be in equilibrium.

The procedure is to make a complete imaginary cut through the truss and, considering the forces **only to one side of the cut**, use the equilibrium equations to solve for the unknown forces in the cut members.

Consider the truss in *example 5.2* and suppose only the forces in members AC, BC and BD are required (*figure 5.9a*).

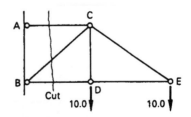

Figure 5.9a
The 'cut' through the structure

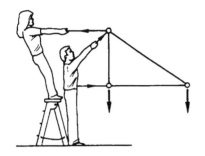

Figure 5.9b

The equilibrium of the structure to one side of the 'cut'

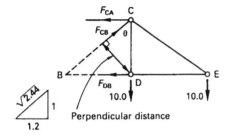

Figure 5.9c

The calculation of unknown forces by the method of sections

A 'cut' is made through the structure which must pass through the member whose force is required. In this case all the required members can be cut at one go; however, sometimes several cuts may be required. The forces on the structure to the left of the cut are completely ignored. The forces in the three cut members must be sufficient to put the remaining section of the structure into equilibrium (*figures 5.9b and c*).

The three basic equilibrium equations can now be used to solve for the three unknown forces. By applying these intelligently it is possible to avoid having to solve simultaneous equations.

To find F_{CA} it is obviously advantageous to eliminate the other two unknown forces F_{CB} and F_{DB}. Therefore take moments about B – the point of intersection of the other two unknowns.

(ΣM about B)
$$(10 \times 1.2) + (10 \times 2.7) = F_{CA} \times 1$$
$$\boldsymbol{F_{CA} = 39 \text{ kN tension}}$$

Likewise, to find F_{DB} take moments about C in order to eliminate F_{CA} and F_{CB}.

(ΣM about C) $10 \times 1.5 = F_{DB} \times 1$
$$\boldsymbol{F_{DB} = 15\text{kN compression}}$$

To find F_{CB} the above strategy cannot operate as F_{CA} and F_{DB} are parallel, however, there is the simple alternative of using vertical equilibrium to eliminate F_{CA} and F_{DB}.

($\Sigma V = 0$) $$\frac{F_{CB} \times 1}{\sqrt{2.44}} = 10 + 10$$

$$\boldsymbol{F_{CB} = 31.2 \text{ kN compression}}$$

Alternatively, it is possible to take moments about D as F_{CA} is now known. This does, however, require the perpendicular distance from D to F_{CB} to be calculated or scaled.

$$\theta = \tan^{-1}(1.2/1) = 50.19°$$

$$\text{Perpendicular distance} = 1 \times \sin 50.19° = 0.768 \text{ m}$$

(ΣM about D)
$$(F_{CB} \times 0.768) + (10 \times 1.5) = F_{CA} \times 1$$

$$F_{CB} = \frac{(39.0 \times 1) - 15}{0.768}$$

$$F_{CB} = \textbf{31.25 kN compression}$$ check

A negative value for a force simply means that the direction was originally chosen wrongly.

The basic strategy to adopt therefore is:

To eliminate the effect of two parallel members – sum forces perpendicular to them.

To eliminate the effect of two members whose lines of action intersect – take moments about the intersection point.

Example 5.5

Figure 5.10a shows a water tower structure with wind blowing on the side of the water tank. Find the force in members AC, BC and CD due to the wind load only. Wind pressure is 0.6 kN/m² and a partial safety factor for load of 1.2 should be used. Scale drawing is acceptable.

Solution

Loads

$$\begin{aligned}
\text{Front area of tank} &= 2 \times 2 &&= 4 \text{ m}^2 \\
\text{Characteristic wind force} &= 4 \times 0.6 &&= 2.4 \text{ kN} \\
\text{Design load} &= 1.2 \times 2.4 = 2.88 \text{ kN}
\end{aligned}$$

Divide between the two sides of the tower:

$$\text{Design load per truss} = \frac{2.88}{2} = \textbf{1.44 kN}$$

This load is assumed to be acting at the level of the centre of the tank, which is 7 m above ground level.

Cut 1 (refer to *figure 5.10b*)

To find F_{CA} eliminate F_{CB} and F_{DB} by taking moments about B.

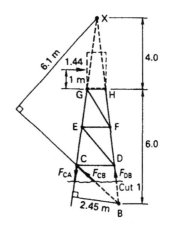

Figure 5.10a
A water tower structure

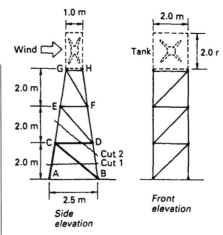

Figure 5.10b
Elimination of forces F_{CB} and F_{DB}

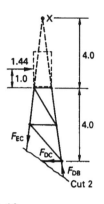

Figure 5.10c
Elimination of forces F_{CA} and F_{DB}

$(\sum M \text{ about B})$ $\quad\quad$ $1.44 \times 7 = F_{CA} \times 2.45$

Answer $\quad\quad\quad\quad\quad$ $F_{CA} = 4.11 \text{ kN tension}$

To find F_{CB} eliminate F_{CA} and F_{DB} by taking moments about point of intersection at X.

$(\sum M \text{ about X})$ $\quad\quad$ $F_{CB} \times 6.1 = 1.44 \times 3$

Answer $\quad\quad\quad\quad\quad$ $F_{CB} = 0.71 \text{ kN compression}$

Cut 2 (refer to *figure 5.10c*)

Again take moments about point X:
$(\sum M \text{ about X})$ $\quad\quad$ $F_{DC} \times 8 = 1.44 \times 3$

Answer $\quad\quad\quad\quad\quad$ $F_{DC} = 0.54 \text{ kN}$

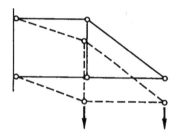

Figure 5.11a
The effect of removing a diagonal member

5.4 Unstable structures and redundant members

5.4.1 Identification by inspection

All of the trusses considered up to now have contained just sufficient members and supports to support the loads. These are known as **statically determinate** structures, and are always capable of analysis by the methods previously described. In *example 5.1* it was shown that if any member is cut or removed, the truss becomes unstable and turns into a **mechanism**. Likewise if either of the pinned supports is reduced to a roller support then a mechanism would result. It is obviously vital to be able to distinguish between structures and mechanisms.

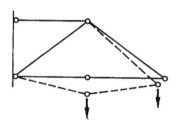

Figure 5.11b
The effect of removing a vertical member

Figure 5.11a shows the structure used in *example 5.1*. However, a diagonal bracing member has been removed. Unbraced quadrilaterals should always be treated cautiously as they indicate a mechanism.

The removal of any member results in collapse (*figure 5.11b*).

Figure 5.11c shows the effect on a structure of reducing the support restraints on a structure.

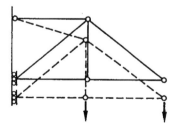

Figure 5.11c
The effect of reducing a pinned support to a roller

If more members are present in a structure than are necessary for equilibrium, then it is known as **statically indeterminate**. This means that it is **not** possible to analyse it fully by the simple methods of equilibrium previously described. In practice, it means that it may be possible to analyse part of the structure by these methods, but sooner or later you would encounter no remaining joint with only two unknowns and the analysis would grind to a halt. Each extra member or support restraint, above the minimum sufficient, is called a **redundancy**.

In *figure 5.12a* it is clear that if either of the diagonal members were removed the structure would not collapse. This structure has one redundancy.

Figure 5.12a
One redundant member

As the structure is loaded one diagonal would go into tension and the other into compression.

   If one of the diagonals was very robust and stiff, and the other very slender and flexible (*figure 5.12b*), then clearly the robust member would carry most of the load and the slender one very little. Thus without information concerning the stiffness of the members the distribution of forces throughout the structure cannot be determined. It is possible to analyse statically indeterminate structures but this is beyond the scope of this book.

Figure 5.12b
The effect of a very stiff redundant member

Figure 5.12c shows a statically indeterminate structure. Crossed members often indicate a redundancy.

Figure 5.12d also shows a structure with one redundancy.

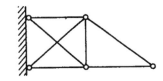

Figure 5.12c
One redundancy

Figure 5.12e shows a structure with four redundancies. One arises from the crossed members and the three others from extra support restraints. A roller adds one and a pin adds two.

It is obviously important to be able to recognise statically indeterminate structures so that time is not wasted in trying to analyse them by simple statics. Statically indeterminate structures can be subject to large changes in member forces if a support settles or a member is not quite the correct length and has to be forced in – known as **lack of fit**. A statically determinate structure, however, can accommodate small errors by changing its geometry.

   Most mechanisms and indeterminate structures can be identified by simple inspection. Try to imagine building the truss member-by-member starting from the supports. If the members were slightly the wrong length, would they have to be forced to fit? If so, the structure is indeterminate. Is the final result stable? If not, we have a mechanism.

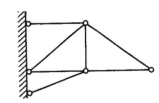

Figure 5.12d
One redundancy

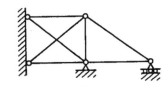

Figure 5.12e
Four redundancies

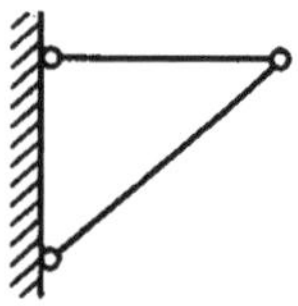

Figure 5.13a
The triangle is the basic building block of trusses

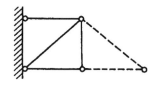

Figure 5.13b
Building up a statically determinate structure

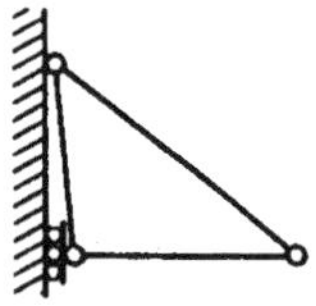

Figure 5.13c
An extra member is required to restrain the roller support

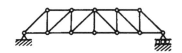

Figure 5.14a
A girder bridge

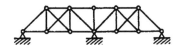

Figure 5.14b
A girder bridge

5.4.2 Identification by application of a formula

Starting with two pinned supports it can be seen that a statically determinate truss is formed by adding two members plus one more pinned joint (*figure 5.13a*). This process can be repeated – each time adding two members plus one joint – and the structure remains statically determinate (*figure 5.13b*).

Hence for a statically determinate structure

$$m = 2j$$

where:

m = number of members
j = number of joints excluding original pin supports.

It follows that if:

$m > 2j$ the structure is indeterminate
$m < 2j$ the structure is a mechanism

If one of the supports is reduced from a pin to a roller, then an extra member is required to restrain the roller (*figure 5.13c*).

Now $m = 2j + 1$ for a statically determinate structure with one roller support

Warning!   Using this formula can be misleading and it is not a substitute for common sense. The following example contains a case where the formula gets it wrong.

Example 5.6

For each of the following pin-jointed trusses (*figure 5.14a–h*), indicate whether the structure is statically determinate, statically indeterminate or a mechanism. In the case of a statically indeterminate structure state the number of redundancies. Assume that a load may be applied at any node.

Solution

(a) By inspection no member could be removed without collapse, therefore the truss is **statically determinate**.

Using the formula: $m = 21$
$j = 10$
Therefore $m = 2j + 1$

One support is a roller therefore the formula checks.

(*b*) By inspection one member can be removed from each of the bays where the members cross. One pinned support, which provides two restraints, can be removed completely. One remaining support can be reduced to a roller. Hence the structure is **statically indeterminate** with **five redundancies**.

Using the formula: $m = 23$
$j = 9$
Therefore $m = 2j + 5$

Formula checks.

(*c*) Crossed members indicate an indeterminacy, but the unbraced quadrilateral is obviously not stable. Therefore the truss is **part statically indeterminate** with one redundancy plus **part mechanism**.

Using the formula: $m = 16$
$j = 8$
Therefore $m = 2j$

The formula indicates a statically determinate structure! This is where the formula approach breaks down and gives a misleading result. The structure has the correct number of members but they are in the wrong place.

(*d*) Either the bottom member could be removed or one support could be made into a roller. Therefore, it is **statically indeterminate** with **one redundancy**.

Using the formula: $m = 19$
$j = 9$
Therefore $m = 2j + 1$

Formula checks.

(*e*) We can remove one crossed member plus one of the members connected to the support without collapse. Therefore the structure is **statically indeterminate** with two **redundancies**.

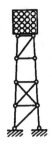

Figure 5.14c
A floodlight tower

Figure 5.14d
A transmission tower

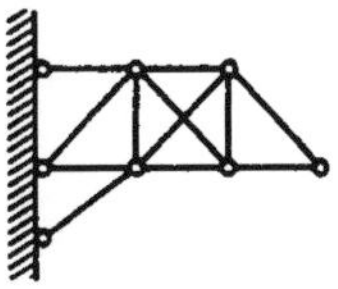

Figure 5.14e
A cantilever bracket

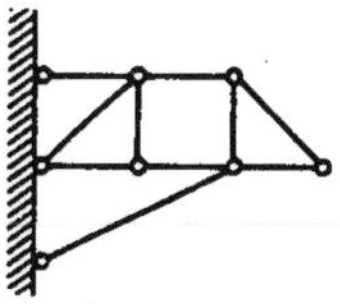

Figure 5.14f
A cantilever bracket

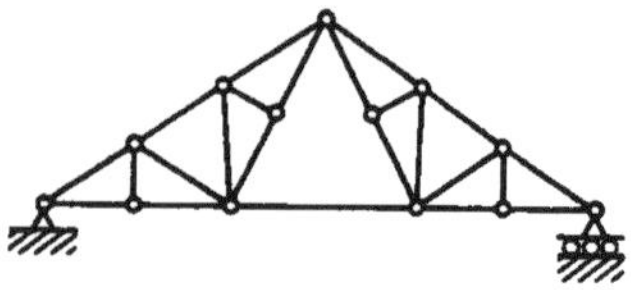

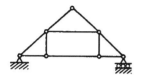

Figure 5.14g
A roof truss

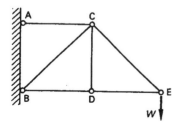

Figure 5.14h
A roof truss

Using the formula: $m = 12$
                     $j = 5$
Therefore $m = 2j + 2$

Formula checks.

(*f*) Not an easy one, but in fact this is **statically determinate**. The extra support member compensates for the unbraced quadrilateral.

Using the formula: $m = 10$
                     $j = 5$
Therefore $m = 2j$

Formula checks.

(*g*) This is a compound roof truss and is **statically determinate** if one support is a roller.

Using the formula: $m = 23$
                     $j = 11$
Therefore $m = 2j + 1$

Formula checks.

(*h*) This is a mechanism because of the unbraced quadrilateral. (In practice some traditional timber roofs use this shape of truss but they are not pin-jointed. The bottom tie member is deep, continuous and hence in bending.)

Using the formula: $m = 10$
                     $j = 5$
Therefore $m = 2j$

But one support is a roller, therefore the formula checks.

Figure 5.15a
Unloaded member, CD, at joint D

5.5 Unloaded members

There are often members in trusses which are unloaded, but they must not be confused with redundant members in statically indeterminate structures. It may be that they become loaded under different conditions.

In *figure 5.15a* it can be seen that, by considering vertical equilibrium of joint D, member CD must be unloaded. However, the fact that it is unloaded does not mean that it is superfluous and can be removed. It would of course be loaded if a load were applied at D. It is still required to restrain joint D and prevent a mechanism forming. It therefore will, in practice, have a small but unknown force in it. It is also significant that members BD and DE are in compression. We will see in *chapter 8* that the load capacity of compression members is dependent on their length, with short members able to carry more load. We can conclude therefore that if member CD were removed and member BDE made continuous it would result in a bigger and more expensive member being required for BDE.

In *figure 5.15b* we can see that, by resolving perpendicularly to either of the members meeting at node E, the forces in members EC and EF are zero. So long as there is not a load applied at E, then both these members are superfluous and can be removed. In general we can say that, **where only two members meet at an unloaded joint, they have zero load**.

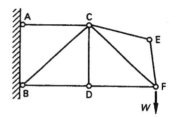

Figure 5.15b
Unloaded members, CE and EF, at joint E

5.6 Crossed tension members

In some structures members are subject to stress reversal. Consider for example the water tower shown in *figure 5.16a*. When the wind blows from the left as shown, the diagonal bracing is in tension. However, if the wind blew from the right the diagonal bracing would be in compression. It must therefore be designed to carry both tensile and compressive loads. Compression members are invariably heavier and hence more expensive than tension members, and so it can sometimes pay to provide double the number of tension members.

Figure 5.16b shows such a structure. This appears to be statically indeterminate but the diagonal bracing members are so slender that they can only operate in tension. If subjected to a compressive load they would simply buckle under negligible load and play no further part in supporting the structure. Thus when the wind is blowing from the left only the bracing shown in *figure 5.16c* operates. When the wind is blowing from the right only the bracing shown in *figure 5.16d* operates. Under this system the bracing could take the form of flat bar or flexible steel cable.

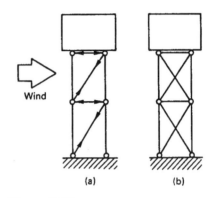

(a)   (b)

Figure 5.16a
Water tower with single bracing acting in either tension or compression

Figure 5.16b
Water tower with crossed bracing

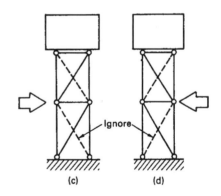

(c)   (d)

Figure 5.16c
Only the bracing shown solid acts when the wind blows from the left

Figure 5.16d
Only the bracing shown solid acts when the wind blows from the right

5.7 Space trusses

Up to now we have been dealing with plane trusses where both the structural members and the loads exist in the same two dimensional

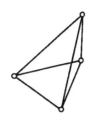

Figure 5.17a
The tetrahedron

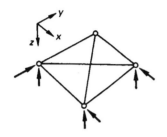

Figure 5.17b
Six reactive forces are required for equilibrium

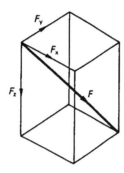

Figure 5.17c
Forces are resolved in three directions

plane. **Space trusses** extend the concept into three dimensions. A space truss is a truly three-dimensional structure. Space trusses are efficient structures, partly because all the joints are well braced, but the connections are often complex and expensive. In this book we shall only be considering very simple space trusses because larger structures invariably require a computer for their design.

We have seen that the basic building block of the plane truss is the triangle, with two members plus one joint being added to extend a truss. The equivalent building block for the space truss is the **tetrahedron** (*figure 5.17a*), with **three** members plus one joint being added to extend structures and maintain statical determinacy.

5.7.1 *Space truss reactions*

We saw that a plane truss required a minimum of three reactive forces to keep it in equilibrium. The equivalent number for a space truss is **six** and this can be seen by considering the structure shown in *figure 5.17b*. Initially imagine that the truss is supported on three roller-ball bearings at A, B and C. This prevents it moving in the Z direction, but it could still slide around on a flat surface in the X–Y plane and so three more reactive forces are required to restrain it. These three reactions must prevent it moving in the X direction, the Y direction and also from rotation.

5.7.2 *Forces in space trusses*

It is possible to analyse space trusses by the method of joints previously described for plane trusses, however, forces must be resolved in **three** directions (*figure 5.17c*) and for large structures this can often lead to having to solve many sets of three simultaneous equations. This is somewhat tedious for hand calculations and the use of a computer is advisable. It is, however, relatively easy to identify certain unloaded members:

Rule 1   If only three members meet at an unloaded joint and they do not all lie in the same plane then they must all be unloaded.

Rule 2   If all but one of the members meeting at a joint are in one plane then the odd member must be unloaded.

For simple structures the intelligent use of the method of sections and the method of joints can save on tedious calculations. This is illustrated in the following example.

Example 5.7

Figure 5.18a shows the plan and elevation of a hoist structure which is attempting to lift a load which does not lie vertically below the point E. Determine the forces in all the members of the structure when the load in the lifting cable is 100 kN.

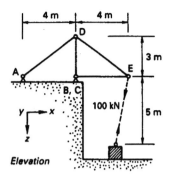

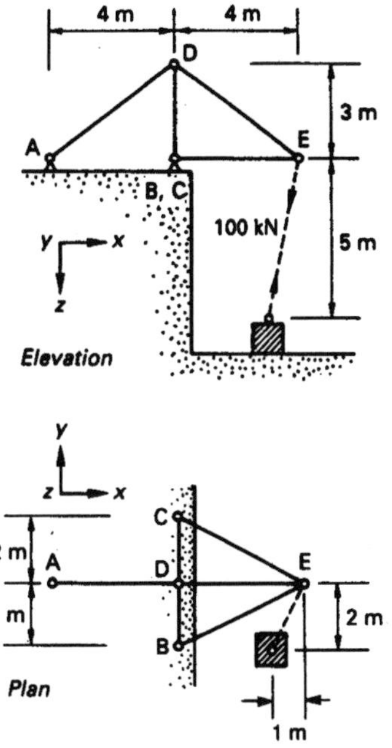

Solution

The first step is to resolve the 100 kN cable force into X, Y and Z components (*figure 5.18b*).

$$l_{cable} = \sqrt{(5^2 + 2^2 + 1^2)} = \sqrt{30}$$

$$= 5.48 \text{ m}$$

$$F_X = \frac{-100 \times 1}{5.48} = -18.3 \text{ kN}$$

Figure 5.18a
Elevation and plan of hoist

$$F_Y = \frac{-100 \times 2}{5.48} = -36.5 \text{ kN}$$

$$F_Z = \frac{+100 \times 5}{5.48} = 91.3 \text{ kN}$$

To find the force in DA use the method of sections (*figure 5.18c*).

To find the perpendicular distance p

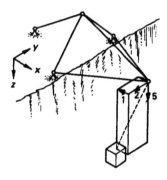

$$\text{Angle } \theta = \tan^{-1} \frac{3}{4} = 36.9°$$

$$p = 3 \cos \theta = 2.4 \text{ m}$$

Figure 5.18b
Resolving the load into X, Y and Z components

(ΣM about B, C)
$$91.3 \times 4 = F_{DA} \times 2.4$$

Answer $F_{DA} = 152.2 \text{ kN tension}$

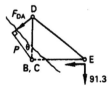

Joint D – Consider the equilibrium of joint D in the X direction.

($\Sigma X = 0$)   **Answer** $F_{DE} = 152.2 \text{ kN tension}$

Figure 5.18c
The method of sections

By symmetry members DB and DC must be equally loaded.

$$l_{DB} = l_{DC} = \sqrt{(3^2 + 2^2)} = 3.61 \text{ m}$$
$$l_{DA} = l_{DE} = 5 \text{ m (3:4:5 triangle)}$$

$(\Sigma Z = 0)$ $\dfrac{2 \times 152.2 \times 3}{5} = \dfrac{2 \times F_{DB} \times 3}{3.61}$

Answer $F_{DB} = F_{DC} = 1109.9 \text{ kN compression}$

Joint E

$$l_{EB} = l_{EC} = \sqrt{(4^2 + 2^2)} = 4.47 \text{ m}$$

$(\Sigma Y = 0)$ $F_{EC} \times \dfrac{2}{4.47} + 36.5 = \dfrac{F_{EB} \times 2}{4.47}$

rearranging $F_{EB} - F_{EC} = 81.6$ [5.3]

$(EX = 0)$ $\dfrac{(F_{EC} + F_{EB}) \times 4}{4.47} = \dfrac{152.2 \times 4}{5} + 18.3$

rearranging $F_{EB} + F_{EC} = 156.5$ [5.4]

solving the two simultaneous equations gives:

Answer $F_{EB} = 119.1 \text{ kN compression}$

Answer $F_{EC} = 37.5 \text{ kN compression}$

5.8 Summary of key points from chapter 5

1. Triangulated or 'trussed' structures are often assumed to have simple pinned joints for design purposes.
2. Provided that loads are only applied at the **nodes** (joints), the structural members can be considered to be in either pure tension or pure compression (i.e. no bending).
3. The **method of joints** is a way of evaluating the forces in the members by considering the equilibrium of each joint in turn.
4. A **composite force diagram** using **Bow's notation** is a more streamlined version of the above.

5. The **method of sections** looks at the equilibrium of 'chunks' of the structure and is a good method if the forces in only a few members are required.

6. The above methods can only be applied to **statically determinate structures**. Structures with too few members or supports are unstable **mechanisms**. Structures with too many members or supports are **redundant** and can only be analysed by more advanced techniques.

7. When only two members meet at an unloaded joint they each have zero load.

8. **Space trusses** have diagonal members in three dimensions, but the analysis of large space trusses can be tedious without a computer.

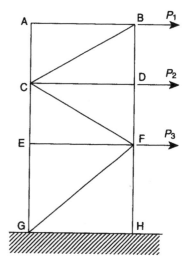

5.9 Exercises

E5.1   For the pin-jointed structure, shown in *figure 5.19* indicate by inspection whether each member is in tension →←
compression —←—→— or unloaded —0—.
tension – BC, CD, CE, EG, FG
compression – BD, CF, DF, FH
unloaded – AB, AC, EF

Figure 5.19
Pin-jointed tower

E5.2   *Figure 5.20* shows a plane pin-jointed structure. Find the forces in all the members and determine if they are tensile or compressive.

Member	Force (kN)	
	tension	compression
AB	74	
BC	56	
CD	102	
EF		105
FG		110
AG		42
BG		53
CG	83	
CE	27	
CF		34

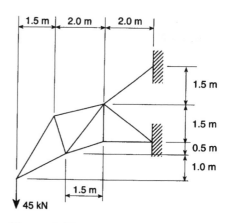

Figure 5.20
Pin-jointed cantilever

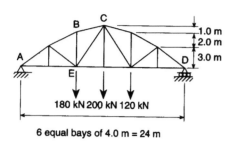

180 kN 200 kN 120 kN

6 equal bays of 4.0 m = 24 m

Figure 5.21
Pin-jointed bridge

E5.3   A pin-jointed bridge structure is shown in *figure 5.21*. Find the forces in members BC and CE only due to the loading given.

F_{BC} = *429 kN tension*, F_{CE} = *28.8 kN compression*

E5.4   For each of the pin-jointed trusses shown in *figure 5.22a–e* indicate whether the structure is statically determinate, *s*, statically indeterminate, *I*, or a mechanism, *m*. In the case of a statically indeterminate structure state the number of redundancies.

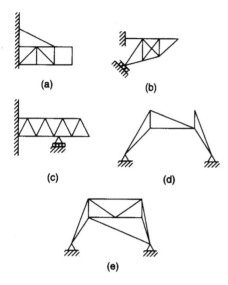

(a)

(b)

(c)

(d)

(e)

Figure 5.22
Pin-jointed trusses

Structure	
a	*i (1)*
b	*m & i (1)*
c	*i (1)*
d	*s*
e	*i (1)*

Chapter 6

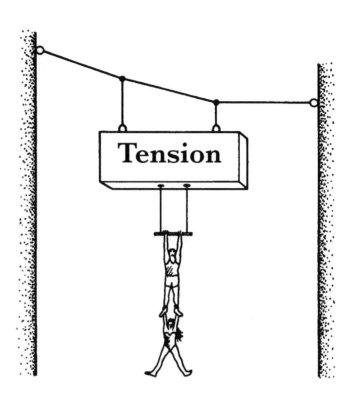

Topics covered

Examples of tensile structures
The design of tension members

Cable structures:
    Light cables with point loads
    Suspension bridges

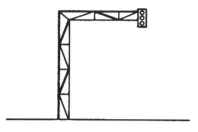

Figure 6.1a
A structure for supporting traffic lights with tensile members shown in bold

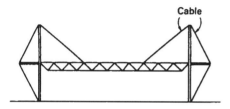

Figure 6.1b
A roof structure supported by tensile cables

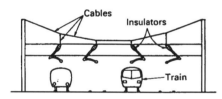

Figure 6.1c
A cable structure supporting overhead electric railway lines

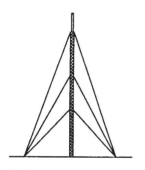

Figure 6.1d
A broadcasting mast supported by cable guys

6.1 Introduction

We saw in the previous chapter how framed structures with pinned joints contain members which are in either pure tension or pure compression, provided loads are only applied at the nodes. We also saw how such structures can be analysed to determine the magnitude of the force in each member. This chapter considers the **design** of members in pure tension, i.e. for a known tensile load, what are the actual required dimensions of a steel or aluminium tie to safely support the load?

We then go on to consider the design of cable structures. These are tensile structures that can be used in highly exciting and innovative ways. Tensile members can be very slender and cables can form graceful curves. A common example is the suspension bridge, which most people invariably regard as a particularly pleasing structural form.

6.2 Examples of tensile structures

We saw in *chapter 5* that many of the structural members of pin-jointed frames are in tension. In *figure 6.1a* the tensile members are indicated by bold lines.

Many interesting 'hi-tech' architectural designs use tensile cables to support floor and roof structures. This permits large clear spans for lightweight structures. *Figure 6.1b* shows the construction for a leisure complex.

Figure 6.1c shows a common tensile cable structure for supporting overhead electric transmission lines for a high speed railway.

Many tall broadcasting masts use tensile guys or stays to improve stability – particularly to counter the effect of lateral load from wind (*figure 6.1d*). The masts of sailing yachts use cables in similar ways.

The cable-stayed bridge shown in *figure 6.1e* uses straight cables or rods to reduce the bending in the bridge deck.

Suspension bridges (*figure 6.1f*) use tensile cables both for the main curved suspension cables and for the hangers that support the deck. Bridges of this type are used for the world's longest spans.

Figure 6.1g shows one of the oldest and most well-known forms of tensile structure.

6.3 The design of axially loaded ties

Ties (members in tension) are the simplest of all members to design. The load required to break a tie is directly proportional to its cross-sectional area. This is the hatched area shown in *figure 6.2*. The aim, therefore, is simply to make the cross-sectional area of the tie big enough to ensure that the stress in it always remains less than a specified safe value.

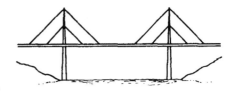

Figure 6.1e
A cable-stayed bridge

From *chapter 2*
$$\text{stress} = \frac{\text{force}}{\text{area}}$$

therefore
$$\text{area} >= \frac{\text{force}}{\text{stress}}$$

(The symbol >= means 'greater than or equal to')

In limit state terms this can be re-written as follows:

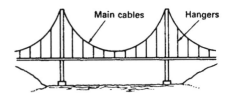

Figure 6.1f
A suspension bridge

$$\textbf{effective area, A}_{\textbf{e}} >= \frac{\textbf{design load}}{\textbf{design strength}}$$

where the **effective area**, $\mathbf{A_e}$, is simply the cross-sectional area of the tie with a suitable reduction made for any bolt holes that might reduce its strength.

The **design strength**, in limit state design, is the characteristic strength divided by γ_m, the partial safety factor for material strength. (*Table 3.6* indicates that γ_m for steel is, in fact, 1.0.)

The **design load** on the tie is the load obtained from the analysis of the structure, remembering that the characteristic loads must be multiplied by γ_f, the partial safety factor for loads given in *table 4.4*.

Figure 6.1g
A washing line is a common tensile structure

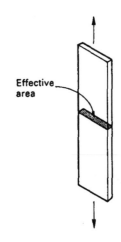

Example 6.1

A tie member in a pin-jointed frame structure must support a design load of 72 kN tension. Determine the required width of 6 mm thick flat mild steel bar.

Solution

$$\text{Minimum effective area} = \frac{\text{design load}}{\text{design strength}}$$

Figure 6.2
The effective area of a plain axially loaded tie

Table 3.4 gives the yield stress of mild steel as 275 N/mm² and *table 3.6* indicates a partial safety factor for material strength of 1.0.

$$\text{Hence minimum A}_\text{e} = \frac{72 \times 10^3}{275} = 262 \text{ mm}^2$$

$$\text{Required width} = \frac{262}{6} = 44 \text{ mm, say 50 mm}$$

Answer   Use 50 mm × 6 mm mild steel bar

6.3.1 *Ties containing holes*

In the above example, the calculated area would be acceptable for a tie in a welded frame where it was not weakened by holes at the ends to accommodate bolts. If bolts are to be used, then generally, the size of the tie must be increased to compensate for the weakening effect. The most likely place for the tie to break is at the hole, where the cross-sectional area is a minimum. *Figure 6.3* shows a typical bolted connection together with the effective area that must be used in calculations. The diameter of the bolt hole must be deducted from the width of the tie.

effective area = (width − bolt diameter) × thickness

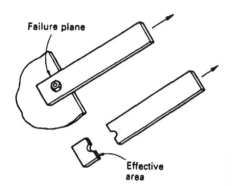

Failure plane

Effective area

Figure 6.3

The effective area of a bolted connection

Example 6.2

Figure 6.4a shows part of a tower crane where the counterweight is supported by an inclined tie as shown. The characteristic dead weight of the counterweight is 150 kN. Ignore the self-weight of the crane structure and any other loads that may be acting.

Determine:

a. The design tensile force in the tie.
b. The required width of a 20 mm thick mild steel tie if it is to contain a single row of 27 mm diameter bolt holes.

Solution

Loads

The first step is to convert the characteristic load into a design load. From *table 4.4* the partial safety factor for dead load is 1.4.

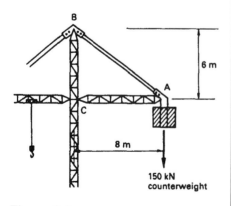

B

6 m

A

C

8 m

150 kN counterweight

Figure 6.4a

Part of a tower crane

Therefore   Design load $= 150 \times 1.4 = 210$ kN

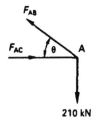

Figure 6.4b
Resolving forces at A

Analysis

We can now determine the force in the tie by using the method of joints at joint A (*figure 6.4b*).

$(\Sigma V = 0)$ $F_{AB} \times \sin \theta = 210$

We can see that ABC is a 3:4:5 triangle

$$F_{AB} = \frac{210 \times 5}{3} = 350 \text{ kN}$$

Answer   Design force in tie = 350 kN tension

Design of tie

As in *example 6.1* the design strength of mild steel is 275 N/mm²

$$\text{Effective area required} = \frac{350 \times 10^3}{275} = 1273 \text{ mm}^2$$

$$\text{Required width} = \frac{1273}{20} + 27 = 91 \text{ mm}$$

Answer   **Use 100 × 20 mild steel tie**

6.3.2 *Angles as ties*

Angle sections are commonly used for ties, particularly in framed structures. A problem is that they are usually connected by bolts or welds through only one leg (*figure 6.5*). This means that the load is not applied axially. The eccentricity of the end connection produces a combination of tension and bending in the tie. To compensate for this a simple empirical formula is used to reduce the gross area of the angle:

$$\text{effective area} = \frac{3a_1 a_2}{3a_1 + a_2} + a_1$$

The areas are as defined in *figure 6.6*.

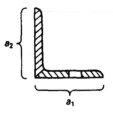

Figure 6.5
An angle connected by only one leg

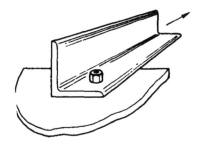

Figure 6.6
The areas used to calculate the effective area of an angle

Example 6.3

Determine the dimensions of a standard mild steel equal angle to support a design load of 210 kN if it contains 22 mm diameter bolt holes in one leg only.

Solution

As in *example 6.1* the design strength of mild steel is 275 N/mm^2

$$\text{Effective area required} = \frac{210 \times 10^3}{275} \ 764 \ \text{mm}^2$$

We now select an angle from the section tables in the *Appendix* whose gross area is somewhat bigger than this area.

Try 90 × 90 × 7 Angle
area = 12.2 cm^2 = 1220 mm^2

$$a_1 = 610 - (6 \times 22) = 478 \ \text{mm}^2$$
$$a_2 = 610 \ \text{mm}^2$$

$$\text{Effective area} = \frac{3 \times 478 \times 610}{(3 \times 478) + 610} + 478$$
$$= 906 \ \text{mm}^2 > 764 \ \text{mm}^2$$

Answer   **Use 90 × 90 × 7 Angle**

6.4 Cable structures

Cable structures can be exciting, lightweight and highly efficient. It is usual to use cables made from a very high grade steel. This produces large concentrations of load, and hence particular care must be taken with the design and manufacture of end connections if catastrophic failure is to be avoided.

A key feature of all design involving cables is that they are assumed to support **only** tensile loads. That is they are assumed to be incapable of carrying compressive forces or bending.

Cable structures can be divided into two main classes. Firstly there are relatively lightweight cables with a few point loads, and secondly there are heavier cables with a more uniform distribution of load. These two cases will now be considered separately.

6.4.1 *Lightweight cables with point loads*

For the purposes of this simplified approach we will actually consider the cables to be weightless relative to the applied point loads. This means that we can assume that the cables are straight between the points of load application.

Firstly consider the simple case of a single point load as shown in *figure 6.7*. Clearly, provided the load is applied in an appropriate direction to keep the cable tight, the geometry is independent of the load and fixed only by the lengths of the cable. Or is it? To be strictly correct, the geometry of the structure changes as the cable extends under load. This is significant for structures such as cable-stayed bridges, but not important for small determinate structures.

The tensile forces in the two sections of the cable can be easily determined by resolving the forces at the load point.

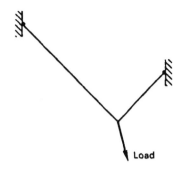

Figure 6.7
A light cable with a single point load

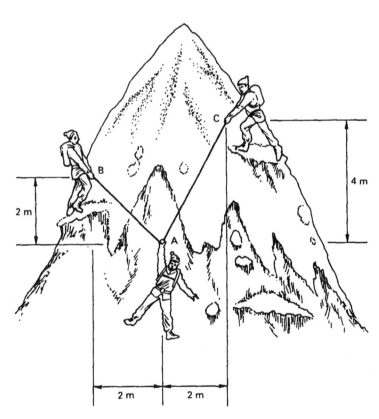

Figure 6.8a
A climber suspended from a cable

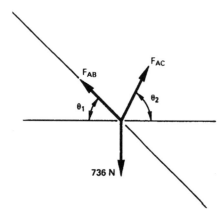

Figure 6.8b
Forces at A

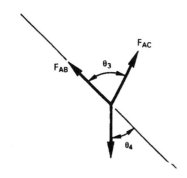

Figure 6.8c
Resolving forces perpendicular to AB

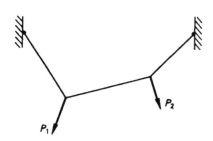

Figure 6.9a
A cable with multiple point loads at various angles

Example 6.4

If the climber suspended from the rope in *figure 6.8a* weighs 75 kg, calculate the force in the rope to each side of him.

Solution

Loads

$$\text{Load from climber} = 75 \times 9.81 = 736 \text{ N}$$

Analysis

From *figure 6.8b*

$$\theta_1 = 45°$$
$$\theta_2 = \tan^{-1} 4/2 = 63.4°$$

Comment – We could now resolve vertical and horizontal forces at A in the usual way, and then solve the resulting two simultaneous equations for F_{AB} and F_{AC}. However, as pointed out in section 5.3.2, if we resolve perpendicular to one of the unknown forces it removes the need to solve simultaneous equations.

From *figure 6.8c*

$$\theta_3 = 180° - 45° - 63.4° = 71.6°$$
$$\theta_4 = 45°$$

(Σforces perpendicular to AB)
$$F_{AC} \times \sin 71.6° = 736 \times \sin 45°$$
Answer $F_{AC} = 548$ **N**

($\Sigma V = 0$) $736 = (F_{AB} \times \sin 45°) + (548 \times \sin 63.4°)$
Answer $F_{AB} = 348$ **kN**

If we consider the general case of multiple point loads acting at various angles as shown in *figure 6.9a*, we are faced with a difficult problem. Not only are the cable tensions unknown, but the basic geometry of the structure changes as the magnitude of the loads change. The structure is effectively a mechanism whose shape changes to restore equilibrium. This is illustrated in *figure 6.9b*. Even without considering the change in length of the cable owing to axial strain, there are too many unknowns for us to analyse such structures using the three basic equilibrium equations. They are therefore beyond the scope of this book.

However, such structures are not very common, and we can carry out useful analysis of simplified structures which contain:

- Only vertical loads
- A partly fixed geometry.

The following example illustrates the power of simple reasoning based on consideration of equilibrium.

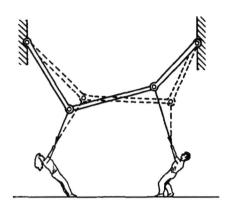

Figure 6.9b
The geometry changes to restore equilibrium as the forces change

Example 6.5

Figure 6.10a shows a cable structure supporting overhead electric railway lines. The loads are as follows:

- Each insulated hanger weighs an estimated 60 kg
- Each hanger supports a conducting line weighing 20 kg
- There is the possibility of an extra imposed load on each conducting line from snow and ice of 400 N.

All other loads, including the self-weight of the supporting cable, may be considered to be negligible.

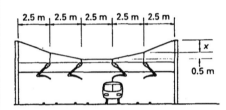

Figure 6.10a
A cable structure supporting overhead electric railway lines

Find:

a. The design point loads.
b. The unknown dimension X.
c. The tensile force throughout the supporting cable.
d. The required diameter of cable. (Assume an ultimate stress of 1570 N/mm^2 and use a partial safety factor for strength, γ_m, of 5.)

Solution

Loads

Design dead load $= 9.81 \times (60 + 20) \times 1.4$	$= 1099$ N
Design imposed load $= 400 \times 1.6$	$= \underline{640}$ N
Total for each point load	$= 1739$ N
say	$= 1.74$ kN

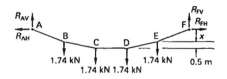

Figure 6.10b
The structural model

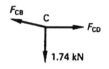

Figure 6.10c
The forces at point C

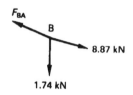

Figure 6.10d
The forces at point B

Analysis

Comment – The lower horizontal cables shown in figure 6.10a provide lateral stability to the insulated hanger, but make no contribution to supporting vertical loads.

The structural model is as shown in *figure 6.10b*. The first observation that we can make is that both the structure and loading are symmetrical, and hence we only need to consider one half.

Point C – See figure 6.10c.

$$l_{BC} = \sqrt{(2.5^2 + 0.5^2)} = 2.55 \text{ m}$$

$(\Sigma V = 0) \qquad F_{CB} \times \dfrac{0.5}{2.55} = 1.74$

Answer $\qquad \mathbf{F_{CB} = 8.87 \text{ kN}}$

$(\Sigma H = 0) \qquad F_{CD} = 8.87 \times \dfrac{2.5}{2.55}$

Answer $\qquad \mathbf{F_{CD} = 8.7 \text{ kN}}$

Point B – See figure 6.10d.

Both the direction and magnitude of F_{BA} are unknown. By considering the horizontal equilibrium of point B we can conclude that the horizontal component of F_{BA} must be equal and opposite to the horizontal component of F_{BC}. Secondly, we can reason that the vertical component of F_{BA} must be double (and opposite) to that of F_{BC} (because an extra downward load of 1.74 kN has been added). These two conditions can **only** be satisfied if the slope of BA is double that of CB. In other words, as extra vertical load is added to the cable its slope must be increased to maintain equilibrium.

Answer dimension, $X = 2 \times 0.5 = \mathbf{1.0 \text{ m}}$

$$l_{AB} = \sqrt{(2.5^2 + 1.0^2)} = 2.69 \text{ m}$$

$(\Sigma H = 0) \qquad F_{BA} \times \dfrac{2.5}{2.69} = 8.87 \times \dfrac{2.5}{2.55}$

Answer $\qquad \mathbf{F_{BA} = 9.36 \text{ kN}}$

Design of cable

$$\text{Design stress} = \frac{1570}{5} = 314 \text{ N/mm}^2$$

$$\text{Area of cable required} = \frac{\text{max. force}}{\text{stress}}$$

$$= \frac{9.36 \times 10^3}{314} \; 29.8 \text{ mm}^2$$

Cable diameter = 6.16 mm

Answer   Use 7 mm diameter cable

Comments:

1. The use of such a large partial safety factor on material strength is justified on the following grounds:

- Structural members with such small cross-sectional areas are particularly susceptible to damage through corrosion or rough handling.
- The consequences of failure are particularly serious.
- Cables are relatively cheap, and so there is only a small increase in cost.

2. Because the cable can support only pure tensile forces, the reactions at the ends must be parallel with the cable. We could of course easily resolve these into vertical and horizontal components.

6.4.2 *Heavy cables with uniform loads*

In the previous section we assumed that the cables formed straight lines between point loads. In this section the cables are considered to be curved.

Figure 6.11 shows a typical centre span of a large suspension bridge. We shall assume that the vertical hangers, which join the bridge deck to the main cables, occur frequently enough to constitute a uniform load on the cable. It is an easy matter to develop expressions for the cable reactions and the force in the cable.

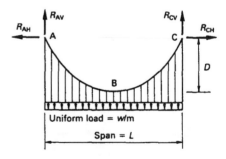

Figure 6.11
The centre span of a suspension bridge

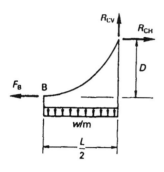

Figure 6.12

The moment at B must be zero

From symmetry and vertical equilibrium we know:

$$R_{AV} = R_{CV} = \frac{wL}{2} \qquad [6.1]$$

Because a cable cannot support bending, the net sum of clockwise and anticlockwise moments to one side of any point must be zero. Consider one half of the span (see *figure 6.12*).

$(\Sigma M \text{ about B}) \qquad \underbrace{(wL/2 \times L/4) + (R_{CH} \times D)}_{\text{clockwise}} = \underbrace{R_{CV} \times L/2}_{\text{anticlockwise}}$

Substitute for R_{CV} from *equation [6.1]* and rearrange:

$$R_{CH} = \frac{wL^2}{8D} \qquad [6.2]$$

We can see, by considering the horizontal equilibrium of the cable in *figure 6.12*, that the force in the cable at mid-span, F_B, is equal to R_{CH}.

The maximum force in the cable occurs at the support and is obtained by finding the resultant of the vertical and horizontal reactions.

$$\text{Maximum force in cable} = \sqrt{(R_{CV}{}^2 + R_{CH}{}^2)} \qquad [6.3]$$

Inspection of *equations [6.2]* and *[6.3]* reveals that the horizontal reaction and the force in the cable is reduced if the height of the bridge, *D*, is increased. Thus tall bridges require thinner cables, but of course the length of the cables and towers is increased.

What is interesting is that the above expressions have been derived without making any assumptions about the actual shape of the cable form. In fact, for a uniform load the cable must be a **parabola**. This is easily proved by showing that, with a parabola, there is no bending at any point in the cable. In effect, the proof is identical to that given in *chapter 12* for the parabolic arch, which is in pure compression, and hence the inverse of this case.

If a cable is supporting only its own self-weight then, because of the change of slope, the weight is not uniformly distributed throughout the length of the cable. Under these circumstances the cable takes up a hyperbolic **catenary** form. For real bridges the form of the curve is obviously complex, and lies somewhere between the straight line, the parabola and the catenary. However, as we shall see in the next example, this is not important in determining the principal member sizes.

Example 6.6

Figure 6.13a shows the Bosphorous Bridge in Turkey. The main cables are made up from a bundle of small wires. The following data applies:

Central span, $L = 1074$ m
Height of cable rise, $D = 95$ m
Concrete anchor blocks $= 42.0$ m $\times$ 20.0 m $\times$ 12.0 m
Slope of cables at towers $=$ same angle both sides of tower

Number of main cables $= 2$
Ultimate strength of cables $= 1590$ N/mm^2
Diameter of wires $= 5$ mm
Number of wires $= 10412$

Dead load (cables + deck), $G_k = 147$ kN/m
Imposed load, $Q_k = 20.3$ kN/m

Determine:

a. The design load for the suspended part of the structure.
b. The total compression load, P, in one supporting tower.
c. The maximum force, T, in the cable.
d. The value of the partial safety factor for material strength, γ_m, in the cable at full load.
e. The factor of safety against vertical uplift of an anchor block.

Solution

Comment – Figure 6.13b shows a visualisation of the forces in a suspension bridge. The Bosphorous Bridge is unusual in having inclined suspension cables for the deck, however, for the purposes of this example they can be assumed to be vertical as shown in the visualisation.

Loads

a. Applying the usual partial safety factors for loads:

$$\text{Design load} = (1.4 \times 147) + (1.6 \times 20.3)$$

Answer   Design load = 238.8 kN/m

Figure 6.13c shows the structural model.

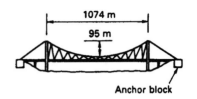

Figure 6.13a
The Bosphorous Bridge, Turkey

Figure 6.13b
Visualisation of suspension bridge with the 'towers' in compression and the cables in tension

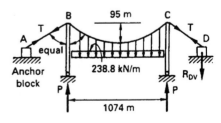

Figure 6.13c
Structural model of suspension bridge

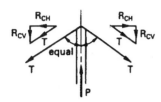

Figure 6.13d
Cable forces at the top of the tower

Analysis

b. The compressive load in a tower is the sum of the vertical components of the force in the cable at each side of the tower – see *figure 6.13d*. If the slope of the cable is the same at each side of the tower, then the vertical component force will be the same at each side of the tower.

From *equation [6.1]*
$$R_{CV} = \frac{wL}{2}$$

$$R_{CV} = \frac{238.8 \times 1074}{2} = 128\ 236\ \text{kN}$$

Answer   **load in tower** $= 2 \times 128\ 236 =$ **256 472 kN**

c. To find the maximum force in the cable we must first find the horizontal component of force.

From *equation [6.2]*
$$R_{CH} = \frac{wL^2}{8D}$$

$$R_{CH} = \frac{238.8 \times 1074^2}{8 \times 95} = 362\ 434\ \text{kN}$$

From *equation [6.3]* – divide by 2 because of two cables.

$$\text{Max. force per cable} = \frac{\sqrt{(R_{CV}^2 + R_{CH}^2)}}{2}$$

$$= \frac{\sqrt{(128\ 236^2 + 362\ 434^2)}}{2}$$

Answer   **Maximum cable force = 192 230 kN**

d. The first step is to find the actual stress in the cable.

$$\text{Area of steel in cable} = \pi \times 2.5^2 \times 10\ 412$$
$$= 204\ 466\ \text{mm}^2$$

$$\text{stress} = \frac{\text{force}}{\text{area}} = \frac{192\ 230 \times 10^3}{204\ 466}$$

$$= 940\ \text{N/mm}^2$$

The partial safety factor for material strength is given by the characteristic ultimate stress divided by the actual stress:

Answer
$$\gamma_m = \frac{1590}{940} = \mathbf{1.69}$$

e. The factor of safety against uplift of an anchor block is the weight of the block divided by the vertical component of force in one cable, R_{DV}.

From *table 4.1* the unit weight of concrete is 24 kN/m^3.

Weight of anchor block $= 42 \times 20 \times 12 \times 24 = 241\,920 \text{ kN}$

$$R_{DV} \text{ in one cable} = \frac{128\,236}{2}\ 64\,118 \text{ kN}$$

Answer

$$\textbf{Factor of safety against uplift} = \frac{241\,920}{64\,118} = \mathbf{3.77}$$

Comment – The Bosphorous Bridge is also unusual in not having suspended side-spans. Figure 6.13e shows the more common arrangement. In this case the anchor blocks only have to support horizontal forces.

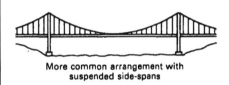

More common arrangement with
suspended side-spans

Figure 6.13e
A more conventional arrangement with suspended side spans

6.5 Summary of key points from chapter 6

1. Tension members (ties) can be relatively slender and cables can be used.
2. The required cross-sectional area of a tie is given by:

$$\text{effective area, } A_e >= \frac{\text{design load}}{\text{design strength}}$$

3. The effective area, A_e must be reduced to allow for bolt holes and eccentric loading on angle members.
4. For a structure consisting of a lightweight cable and relatively heavy loads it can be assumed that the cable forms straight lines between the load points.
5. A structure with a single loading point can be analysed by considering the equilibrium of that point. In most cases where there is more than one loading point the geometry of the structure will change to restore equilibrium. This makes analysis difficult.

6. When a cable is relatively heavy, compared to the loads, it forms a curve as in a suspension bridge.

7. For a suspension bridge cable with uniform load, the horizontal component of force in the cable is given by:

$$R_{CH} = \frac{wL^2}{8D}$$

and the vertical component at the tower is given by:

$$R_{AV} = R_{CV} = \frac{wL}{2}$$

8. The maximum force in the cable is determined by finding the resultant of the above components:

$$\text{Maximum force in cable} = \sqrt{(R_{CV}^2 + R_{CH}^2)}$$

6.6 Exercises

E6.1   A mild steel tie member is required to support a design tensile load of 150 kN. Compare the steel area required of the following two options:

a) A 5 mm thick flat bar with a single row of 22 mm diameter holes for bolts (round-up width to the nearest 5 mm).

b) An equal angle with only one leg of the angle welded to form the connection.

a) *Flat area = 675 mm²*

b) *Angle area = 691 mm²*

E6.2   A lightweight cable is used to support two sets of traffic lights over a road as shown in *figure 6.14*. The design loads can be considered to be two point loads of 2.0 kN each.

a) Determine the force in the cable both at the support and at mid-span.

b) What diameter of cable is required if the material has a yield stress σ_y of 1200 N/mm² and a partial safety factor for material strength, γ_m of 5 is required.

a) *forces = 3.60 kN and 3.00 kN*

b) *cable diameter = 4.4 mm, say 5 mm*

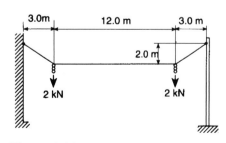

Figure 6.14
Cable-suspended traffic lights

E6.3 *Figure 6.15* shows a suspended footbridge. The following data applies:

Central span   = 64 m
Side-spans   = 32 m
Height of cable rise   = 15 m
Width of walkway   = 2 m
Imposed deck load   = 5 kN/m^2
Dead load (cable and deck) = 6 kN/m
Number of main cables   = 2

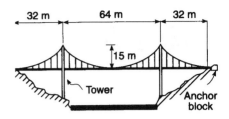

Figure 6.15
Suspended footbridge

Determine:
a) The maximum force in the cable.
b) The compressive force in a tower.
c) The diameter of a suitable cable if material with a yield stress of 1590 N/mm^2 is used combined with a partial safety factor for material strength of 3.
d) The horizontal force acting on one of the cable anchor blocks.

a) *cable force = 571 kN*
b) *tower force = 1562 kN*
c) *cable diameter = 37.0 mm say 40.0 mm*
d) *horizontal anchor force = 417 kN/cable*

E6.4 *Figure 6.16* shows a suspended lighting rail for use in a theatre. Each light weighs 25 kg. All other loads can be ignored. For the particular light positions shown:
a) Determine the design force in the two wire suspension cables if a partial safety factor for loads of 1.6 is used.
b) Assuming a partial factor of safety for material strength of 4, determine the required diameter of steel wire if the material has a yield stress of 300 N/mm^2

a) *forces = 472 N and 704 N*
b) *diameter = 3.86 mm, say 4.0 mm*

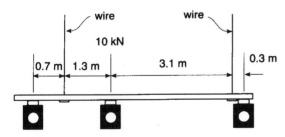

Figure 6.16
Suspended theatre lights

Chapter 7

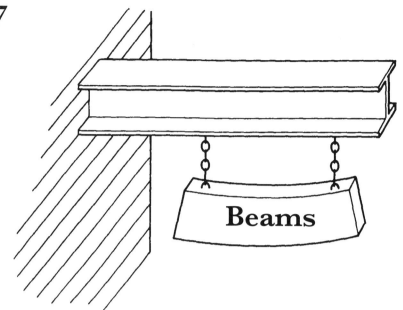

Beams

Topics covered

Examples of beams
The concepts of shear force and bending moment
Drawing shear force and bending moment diagrams
Bending stresses – plastic and elastic
Shear stresses in beams
The design of standard beams in steel, concrete and timber
The calculation of section properties for non-standard beams
Beams without lateral restraint

7.1 Introduction

A **beam** is a structural member subject to **bending** and is probably the most common structural element that designers have to cope with. Bending occurs in a member when a component of load is applied perpendicular to the member axis, and some distance from a support. Bending causes **curvature** of a member. Commonly, beams are horizontal, and loads vertically downwards. Most beams span between two or more fixed points as shown in *figure 7.1a*. The types of support were discussed in *section 2.5*. As with plane trusses there must be at least three reactive forces to put a beam into equilibrium. In *figure 7.1a* the supports are one pin and one roller, and this is called **simply supported**. Another common beam type is the **cantilever**, as shown in *figure 7.1b*. Here one end is unsupported but the other must be rigidly built-in to prevent rotation. The third reaction force has now become a moment or couple, but more of that later. A simply supported beam can have a cantilevered as shown in *figure 7.1c*.

With a few exceptions, beams with extra supports such as that shown in *figure 7.1d* are statically indeterminate and their analysis is beyond the scope of this book. These are known as **continuous beams**. There are too many unknown support reactions for solution by the three basic equilibrium equations.

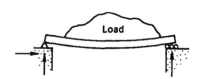

Figure 7.1a
A simply supported beam

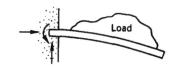

Figure 7.1b
A cantilever beam

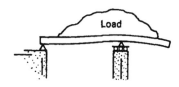

Figure 7.1c
A simply supported beam with a cantilever end

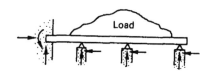

Figure 7.1d
A continuous beam

7.2 Examples of beams and beam types

Figure 7.2a shows the use of a standard **universal beam** in a steel-framed multi-storey building. Universal beams are widely available from local stock holders in a range of sizes and weights. Some of these are tabulated in the *Appendix*. The beam is made up of two flanges and a web – *figure 7.2a*. The 'I' shape of the beam cross-section is efficient, as the steel works more effectively if it is concentrated in the flanges at the top and bottom of the beam. This will be proved later.

Bridges often have a requirement for a long-span beam. A modern form of bridge is the steel box-girder, made by welding flat steel plates together to form a hollow stiffened box (*figure 7.2b*).

Alternatively, multi-storey buildings can have reinforced concrete frames. *Figure 7.2c* shows a typical cross-section

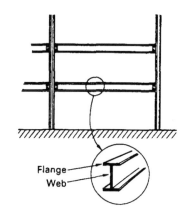

Figure 7.2a
Universal beams in a steel-framed building

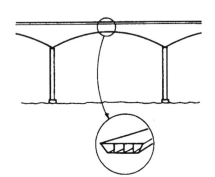

Figure 7.2b
A steel box-girder bridge

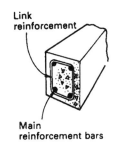

Figure 7.2c
A reinforced concrete beam

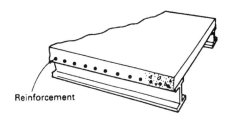

Figure 7.2d
A concrete slab spanning between steel beams

through a reinforced concrete beam. We saw in *chapter 3* that concrete is reasonably strong in compression but weak in tension. In fact we usually ignore the tensile strength of concrete altogether, and provide steel reinforcement to resist the tensile forces. The main reinforcement bars must occur on the tensile side of the beam.

Slabs are widely used for floors and roofs in buildings. They are designed simply as wide but shallow beams. *Figure 7.2d* shows a concrete slab spanning between steel beams. Again the reinforcement must be on the tensile side of the slab.

Extra economy can be achieved by joining a slab to a beam, so that they function as one combined unit. *Figure 7.2e* shows two examples of this. Firstly there is a reinforced concrete 'T' beam, where the slab is cast monolithically with the beam. Secondly there is **composite construction**, where a concrete slab is joined to a steel beam by **shear connectors** welded to the top of the beam at regular intervals.

A variant of the above is **stressed skin construction** where plates are fixed to both the top and bottom of a beam. *Figure 7.2f* shows the aluminium alloy floor of an aircraft, where rivets have been used to join the components together.

Small timber beams are often used in the floor construction of houses, where they are known as **joists** (*figure 7.2g*).

For larger timber beams either **plywood box beams** or **laminated timber beams** can be used (*figure 7.2h*). Laminated beams are formed from strips of timber which are glued together. They can be sanded and varnished to produce a high quality appearance, but they tend to be rather expensive.

7.3 The concept of shear force and bending moment

Stresses within a member due to bending are clearly more complex than those from simple axial tension. An important step in evaluating the stresses within a beam is to determine the magnitude of the shear force and bending moment at the point under consideration. Students often seem to find these concepts difficult, but it is worth persevering with them as, with a little practice, they can be very rewarding.

7.3.1 Cantilever with point load

Consider the cantilever beam shown in *figure 7.3a* which has a point load of magnitude *P* at the end. Make an imaginary cut through the beam at some point distance *x* from the end and consider the equilibrium of the section of beam to the right of the cut (*figure 7.3b*). The sum of the stresses on the cut surface **must** ensure the equilibrium of this piece of beam. Considering the equilibrium of an isolated section of a structure in this way is a common and powerful technique. The section under consideration is known as a **free body**. Think of the free body floating weightlessly in space. When the load is applied, the free body would move if it were not for the presence of stresses at the cut surface. (There is an obvious parallel here with the method of sections analysis of plane frames, where the forces in the cut members must put a section of the structure into equilibrium.)

Clearly a force, *V*, is required to satisfy vertical equilibrium (*figure 7.3c*). In this case obviously *V* = *P* and is independent of the distance along the beam, *x*. This force *V*, which is always normal to the axis of the beam, is termed **shear force**.

We have still not satisfied equilibrium, however, and, as it stands, the free body would spin around in a clockwise direction. To prevent this we need a couple or moment, *M*, applied as shown in *figure 7.3d*. In this case, taking moments about the cut:

$$M = P \times x$$

M is termed the **bending moment** and as expected increases as the distance from the load increases. It reaches a peak value of *M* = *PL* at the wall support. Clearly, in the case of cantilevers, the support must be capable of resisting this moment if collapse is to be avoided.

If values of *V* and *M* are plotted for all values of *x*, to some suitable scale, we get the **shear force diagram** (*figure 7.3e*) and the **bending moment** diagram (*figure 7.3f*). We shall see later that bending moment and shear force diagrams are important because we use them to determine the actual sizes of beams.

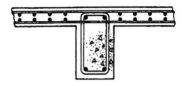
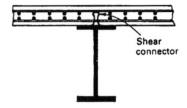

Figure 7.2e
A concrete 'T' beam and composite construction

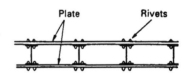

Figure 7.2f
The aluminium floor of an aircraft

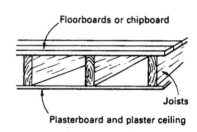

Figure 7.2g
Typical house floor construction using timber joists

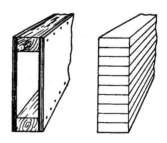

Figure 7.2h
A plywood box beam and a laminated timber frame

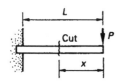

Figure 7.3a
A cantilever beam with a point load

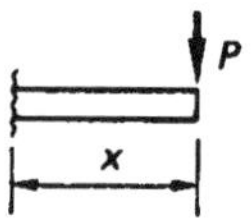

Figure 7.3b
Free body diagram

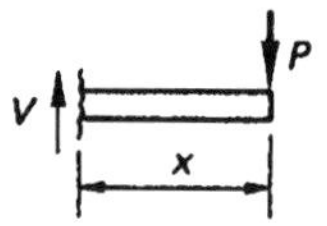

Figure 7.3c
Vertical equilibrium satisfied

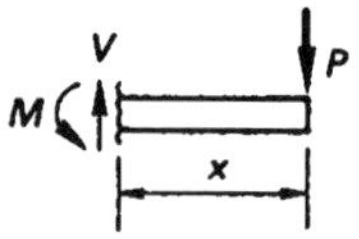

Figure 7.3d
Moment equilibrium also satisfied

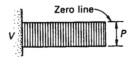

Figure 7.3e
Shear force diagram

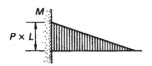

Figure 7.3f
Bending moment diagram

7.3.2 Convention for shear force and bending moment diagrams

It is important to be consistent in the way that shear force and bending moment diagrams are drawn. Different designers use different conventions but in this book the following is used throughout:

Clockwise shears (i.e. when the left-hand side of the beam is being pushed upwards and the right-hand side of the beam is being pushed downwards – *figure 7.4a*) are drawn **below** the zero line.

Anticlockwise shears (*figure 7.4b*) are drawn **above** the zero line.

Bending moments are always drawn on the **tensile side** of the beam.

Therefore:

Hogging moments (*figure 7.4c*) are drawn **above** the line.

Sagging moments (*figure 7.4d*) are drawn **below** the line.

We shall now look at two more commonly occurring cases.

7.3.3 Simply supported beam with point load at mid-span

The beam has a span of L and a point load of P at mid-span (*figure 7.5a*).

The first step is to find the reactions. In this case from symmetry:

$$R_A = R_B = \frac{P}{2}$$

We can now draw the shear force diagram. We will start from R_B and work to the left, considering vertical equilibrium as we go. Inspection of the free body (*figure 7.5b*) indicates an anticlockwise shear which means that the diagram is plotted above the line. However, as we go beyond the mid-span load the direction of the shear force is reversed, and it becomes clockwise (*figure 7.5c*). It can be seen that the value of the shear force is $V = P/2$ throughout in order to satisfy vertical equilibrium. The final shear force diagram is as shown (*figure 7.5d*).

At this stage we can deduce a simple procedure for ensuring that the shear force diagram always complies with our convention:

Start at the right-hand side of the structure and follow the direction of each force as you work towards the left-hand side.

So in this case we start at the right-hand reaction R_B and draw a line vertically upwards representing $P/2$. As we move to the left there no change until we reach the central point load, where we descend an amount P. There is no further change until we reach the reaction R_A where we move up $P/2$. This should always take us back to the baseline and hence provides a check.

Turning to bending moment, it should firstly be obvious that this is a sagging beam with the bottom of the beam in tension. We therefore plot the diagram below the line. The magnitude of the bending moment, M, varies with the distance along the beam, x. Up to the mid-point we can see from *figure 7.5e* that:

$$(x < L/2) \qquad M = \frac{P \times x}{2}$$

Beyond the mid-point we can see from *figure 7.5f* that:

$$(x > L/2) \qquad \begin{aligned} M &= (P/2 \times x) - [P \times (x - L/2)] \\ M &= P \times (L/2 - x/2) \end{aligned}$$

At the mid-span point itself:

$$(x = L/2) \qquad M_{max} = \frac{PL}{4}$$

If these values are now plotted for the whole beam, then the bending moment diagram shown in *figure 7.5g* results.

7.3.4 Simply supported beam with uniformly distributed load

We shall now consider the case of a beam of span L which carries a uniformly distributed load (UDL) of w kN per m (*figure 7.6a*). The dashed line in *figure 7.6a* shows the deflected shape of the beam. Particularly with more complex cases, it is often helpful to

Figure 7.4a
Clockwise shear

Figure 7.4b
Anticlockwise shear

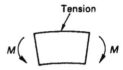

Figure 7.4c
Hogging moment

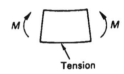

Figure 7.4d
Sagging moment

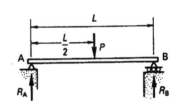

Figure 7.5a
A simply supported beam with a point load at mid-span

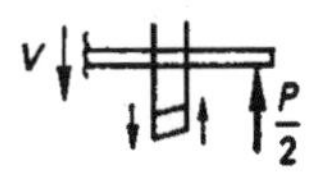

Figure 7.5b
Free body to the right of P

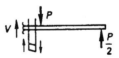

Figure 7.5c
Free body to the left of P

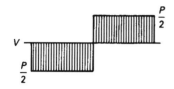

Figure 7.5d
Shear force diagram

Figure 7.5e
Free body to the right of P

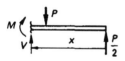

Figure 7.5f
Free body to the left of P

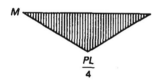

Figure 7.5g
Bending moment diagram

start by drawing the expected deflected shape. The tension side of the beam, on which the bending moment is drawn, is usually obvious.

Shear forces   The free body diagram (*figure 7.6b*) shows how the shear force, V, now varies with x as the gradually applied downward load affects the vertical equilibrium of the free body. The shear force must be equal to the difference between the downward load and the upward reaction. Thus as x increases, V gradually reduces until it is zero at mid-span (where the load and the reaction are equal). Beyond mid-span there is a change of sign. You can obtain the value of V at any point in the beam by simply substituting the appropriate value of x into the following equation:

$$V = \frac{wL}{2} - (w \times x)$$

Plotting this results in the shear force diagram shown in *figure 7.6c*. As expected, the slope of the shear force line is equal to the magnitude of the UDL.

Bending moments   The magnitude of the UDL load is wx. When taking moments we consider the UDL to be concentrated at its mid-point. Remember we always take moments about the cut end of the free body. Therefore all distances are measured from this point when we take moments. The distance from the cut to the mid-point of the UDL is thus $x/2$.

$$M = (wL/2 \times x) - (wx \times x/2)$$

$$M = \frac{w \times (Lx - x^2)}{2}$$

The above is the equation of a parabola and when plotted produces the bending moment diagram shown in *figure 7.6d*.
At mid-span when $x = L/2$ we find:

$$M_{\text{max}} = \frac{wL^2}{8}$$

7.3.5 General rules concerning shear force and bending moment diagrams

In practice, bending moment diagrams are not usually drawn by plotting equations, as suggested above, but by simply evaluating the peak moments at key points and then sketching the intermediate shape. Peak values for common cases, such as $M = PL/4$ for a point load at mid-span and $M = wL^2/8$ for a UDL over the whole span, should be remembered.

We can conclude from the above cases that:

Between point loads   – Shear force is constant.
                           – Bending moment varies as a straight line.

Throughout a UDL   – Shear force varies as a straight line.
                       – Bending moment varies as a parabola.

The peak value of the bending moment occurs at the point in the beam where the shear force passes through zero.

This last point can be very useful for locating the position and magnitude of the maximum bending moment. We shall now prove it mathematically. Those readers who are not interested in the proof can skip the following section.

Proof that a peak value of bending moment occurs at a point of zero shear

Consider a free body consisting of a short length, δx, of beam supporting a UDL of magnitude $w/$m (*figure 7.7*). The values of shear force and bending moment increase from V and M at one end to $V + \delta V$ and $M + \delta M$ at the other.

Now consider the moment equilibrium of the free body by summing moments about point A:

(ΣM about A)

$$M + V\delta x - (w\delta x \times \delta x/2) + M + \delta M$$

Neglecting multiples of small quantities and cancelling M from both sides:

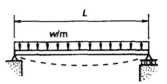

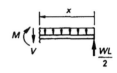

Figure 7.6a
A simply supported beam with a UDL

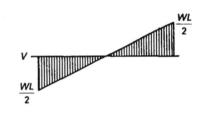

Figure 7.6b
Free body

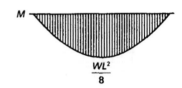

Figure 7.6c
Shear force diagram

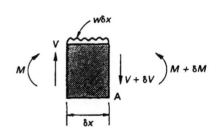

Figure 7.6d
Bending moment diagram

Figure 7.7
A short length of beam

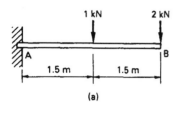

(a)

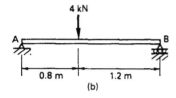

(b)

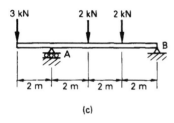

(c)

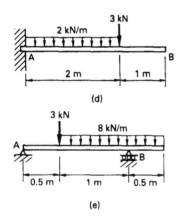

(d)

(e)

Figure 7.8

$$V\delta x = \delta M$$

As δx approaches zero this can be written as the differential:

$$V = \frac{dM}{dx} \qquad [7.1]$$

In words, this means that the shear force is equal to the rate of change of the bending moment diagram. Thus when the shear force is zero the slope of the bending moment is zero, indicating either a maximum or minimum value.

Example 7.1

For each of the beams shown in *figure 7.8a–e* draw the deflected shape, shear force and bending moment diagrams. Indicate the magnitude of all peak values.

Solution

Beam (a)

This is clearly a cantilever beam, and as there is no unknown reaction at the end B we can start straight away with the shear force diagram. Starting at B we draw a line vertically downwards representing 2 kN (*figure 7.9a*). Moving leftwards there is no change in vertical load until we reach the next point load where we descend another 1 kN. Continuing leftwards without further change we find that, at A, we require an upward force of 3 kN to return to the base-line. This is the upward vertical reaction at A, and we can see by inspection that vertical equilibrium is satisfied. i.e. the sum of the downward forces equals the upward reaction.

As far as the bending moment is concerned, because of the point loads, we would expect it to consist of a series of straight lines. We can also see by inspection that these particular loads would cause the beam to deflect downwards at B, producing tension in the top face. We therefore plot the diagram on the top. If we consider the free-body diagram shown, it is clear that the bending moment:

$$M = 2 \times x$$

Hence it has a zero value at B (where $x = 0$) and a value of 3 kNm at the mid-point of the beam. At the support A we must take both point loads into account. Hence:

$$M_A = (2 \times 3) + (1 \times 1.5)$$
$$= 7.5 \text{ kNm}$$

Remember, if we want the moment at A, then all distances must be measured from A.

Comment – This result tells us that both the wall and the beam connection at A must be capable of resisting a moment of 7.5 kNm if collapse is to be avoided.

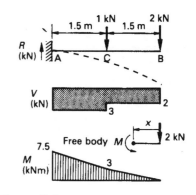

Figure 7.9a
Solution to beam *a*

Beam (b)

In this case we must first find the unknown reaction at B before we can start to draw the shear force diagram.

(ΣM about A) $4 \times 0.8 = R_B \times 2$

                               clockwise   anticlockwise

                    Hence $R_B = 1.6$ kN ↑

We could find R_A in a similar way, by taking moments about B, but there is an easier way. Simply use vertical equilibrium.

($\Sigma V = 0$) $4 = R_A + 1.6$

                     downwards   upwards

Hence $R_A = 2.4$ kN ↑

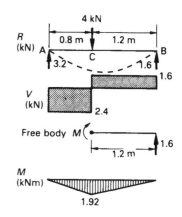

Figure 7.9b
Solution to beam *b*

We are now in a position to draw the shear force diagram. Again start at the right-hand end and follow the direction of the force (*figure 7.9b*).

The peak bending moment will occur at the position of the point load, C, and from the free body, is given by:

$$M_C = 1.2 \times R_B = 1.2 \times 1.6$$
$$= 1.92 \text{ kNm}$$

Clearly we would get the same answer if, as an alternative, we had considered the equilibrium of the left-hand end:

$$M_C = 0.8 \times R_A = 0.8 \times 2.4$$
$$= 1.92 \text{ kNm}$$

This time the bottom of the beam is in tension.

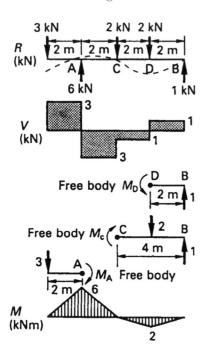

Figure 7.9c
Solution to beam *c*

Beam (c)

First find the reactions.
($\sum W$ about A) $(2 \times 2) + (2 \times 4) = (3 \times 2) + (R_B \times 6)$

                                           clockwise   anticlockwise

$$R_B = 1 \text{ kN} \uparrow$$

($\sum V = 0$) $3 + 2 + 2 = R_A + 1$

                                    downwards   upwards

$$R_A = 6 \text{ kN} \uparrow$$

Comment – Examination of the above calculations shows that if the 3 kN load on the cantilever was doubled, the reaction at B would reduce to zero as the moments became balanced about A. A further increase in the load would result in R_B becoming negative, which implies that the beam must be held down at B in order to maintain equilibrium. Thus when calculating reactions, if a negative sign results, it simply means that the direction of the reaction was originally guessed wrongly.

Hence the shear force diagram can be drawn (*figure 7.9c*)

We can now determine the bending moment at the significant points by breaking the beam, and considering the equilibrium of the resulting free body by summing the moments to one side of the break. We choose either the left-hand or right-hand side depending upon which is the simpler calculation.

$$M_D = -1 \times 2 = -2$$
$$M_C = (-1 \times 4) + (2 \times 2) = 0$$
$$M_A = -3 \times 2 = -6$$

If these values are plotted and joined up by straight lines we have the bending moment diagram.

Comment – In the preceding calculations clockwise moments have been considered positive and anticlockwise moments negative. However, inspection of the bending moment diagram reveals no apparent connection between these signs and the side of the beam on which the bending moment is plotted. This is because we changed from considering the right-hand end to considering the left-hand end. Remember that our convention is always to plot the diagram on the tension side.

Beam (d)

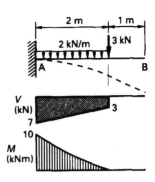

This is a cantilever, similar to beam (a), but with the added complication of a UDL. Starting at B we can see that the first metre is unloaded and hence remains at zero on the shear force diagram (*figure 7.9d*). We then have a 3 kN vertical drop followed by a gradually applied load (at the rate of 2 kN/m), i.e. over 2 m this represents a further downward force of 4 kN.

Because of the UDL, the bending moment will be curved.

$$M_A = (3 \times 2) + (2 \times 2 \times 1)$$
$$= 10 \text{ kNm}$$

Figure 7.9d
Solution to beam *d*

In the above term for the UDL we multiply the magnitude (2 kN/m) by the length over which it acts (2 m) by the distance from A to the centre of the UDL load (1 m). *Figure 7.9d* shows the resulting bending moment diagram.

Comment – It is an easy matter to plot the actual shape of the curve by evaluating the bending moment at several intermediate points. Simply cut the structure at any point, and determine the value of M to put the resulting free body into equilibrium.

Beam (e)

Firstly evaluate the reactions:

(ΣM about A)
$$(3 \times 0.5) + (8 \times 1.5 \times 1.25) = R_B \times 1.5$$
$$R_B = 11 \text{ kN} \uparrow$$

($\Sigma V = 0$)
$$3 + (8 \times 1.5) = R_A + 11$$
$$R_A = 4 \text{ kN} \uparrow$$

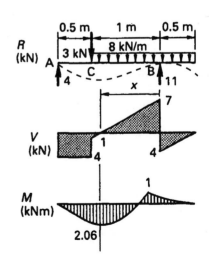

The shear force diagram is shown in *figure 7.9e*. Look at it carefully and make sure that you understand the shape. A significant point is the fact that the diagram crosses the zero axis at two points, and we have already shown how this indicates maxima or minima values of bending moment. We can calculate the distance of the zero shear point from B as follows:

$$\text{Distance, } x = 7/8 = 0.875 \text{ m}$$

Figure 7.9e
Solution to beam *e*

The above is simply the value of the shear force at B divided by the slope of the line (which is equal to the UDL).

For the bending moment, firstly calculate M at B by considering the cantilever end to the right:

$$M_B = 8 \times 0.5 \times 0.25$$
$$= 1 \text{ kNm}$$

Now evaluate the moment at C by considering the left-hand end:

$$M_C = 4 \times 0.5 = 2 \text{ kNm}$$

Lastly evaluate the maximum value of moment which occurs 0.875 m from B:

$$M_{max} = (8 \times 1.375 \times 1.375/2) - (11 \times 0.875)$$
$$= -2.06 \text{ kNm}$$

The final bending moment diagram is shown in *figure 7.9e*.

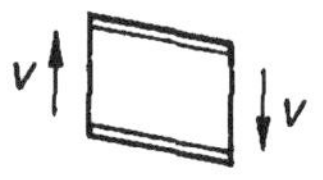

Bending moment diagrams can become quite complicated, however, you should not lose sight of what they mean in physical terms. Quite simply, the shear force and bending moment at any point in a structure are those values necessary to satisfy equilibrium if a free body is considered to one side of the point.

Beams therefore have to resist shear forces which try to deform them as shown in *figure 7.10a*, and bending moments which try to deform them as shown in *figure 7.10b*. We shall consider bending moments first, and the next stage is to understand the relationship between the bending moment, M, on a beam and the bending stresses, σ, that this produces in the beam.

Figure 7.10a
The effect of shear

Figure 7.10b
The effect of bending

7.4 Bending stresses

7.4.1 *Approximate bending strength (ignoring the web)*

An 'I' beam resists bending by one flange 'pushing' and the other flange 'pulling'. The web also contributes to the bending strength, but it is less effective than the flanges. A crude, but safe, approximation to 'I' beam behaviour is to ignore the contribution of the web to bending strength and hence assume that it is the flanges that do all the work. In practice this would not be an economic method of design but it gives a useful insight into the way beams work. In effect, as shown in *figure 7.11a* we are assuming that the web functions in the way that diagonal bracing does in a truss. It keeps the flanges apart and carries the shear.

The moment, M, at section y–y, produces the two flange forces, F_T in tension and F_C in compression as shown in *figure 7.11b*. The distance between the centres of the top and bottom flanges is D. As in the method of sections for pinned trusses:

(ΣM about A) $P \times x = F_T \times D$

but $Px = $ bending moment $= M$

therefore $M = F_T \times D$
in the above   force, $F_T = $ stress $\times$ area

so for a beam with equal flanges of area, A,

$$M = \sigma_t \times A \times D$$
where $\sigma_t = $ the tensile stress in the flange

and if the beam is made from material with an ultimate yield stress of σ_y then the ultimate moment that the beam can support is:

$$M_{ult} = \sigma_y \times D \times A$$

It can clearly be seen that the moment capacity increases with beam depth and flange area. This explains why 'I' shaped beams are used where the aim is to concentrate the material as far apart as possible.

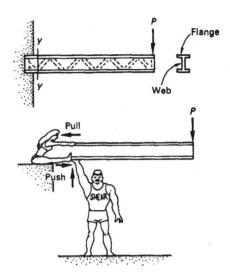

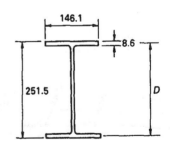

Figure 7.11a
One flange pushes and the other pulls

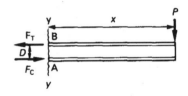

Figure 7.11b
Take moments about the point A

Example 7.2

A 254 × 146 × 31 kg/m universal beam is used as a 3 m long cantilever. What is the maximum ultimate load that can be applied at the end of the cantilever if the steel yields at a stress of 275 N/mm²? Use an approximate method ignoring the contribution of the web, and ignore the self-weight of the beam.

Solution

The first step is to extract the relevant dimensions of the beam cross-section (*figure 7.12a*) from the table given in the *Appendix*.

Figure 7.12a
Beam dimensions

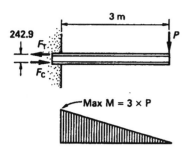

Figure 7.12b
Bending moment diagram

Depth to centre of flanges
$$D = 251.5 - 8.6 = 242.9 \text{ mm}$$

Flange area, $A = 146.1 \times 8.6 = 1256 \text{ mm}^2$

$$M_{ult} = \sigma_y \times D \times A$$
$$= 275 \times 242.9 \times 1256 \times 10^{-6}$$
$$= 83.9 \text{ kNm}$$

Clearly the maximum bending moment occurs adjacent to the support (*figure 7.12b*).

$$\text{Maximum } M = 3 \times P$$

therefore at failure $3 \times P = 83.9 \text{ kNm}$

Answer   **$P = 28.0$ kN**

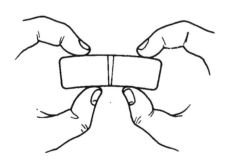

Figure 7.13a
Pencil eraser under no load

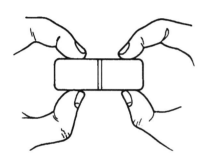

Figure 7.13b
Pencil eraser after bending

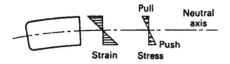

Figure 7.13c
Stress and strain in eraser after bending

7.4.2 Elastic and plastic bending

We shall now develop a more accurate theory for the relationship between bending moments and stresses. The whole cross-section, including the web, is included. Initially the treatment is descriptive but we shall then go on to derive the mathematical relationships.

Consider a pencil eraser with two parallel lines as shown in *figure 7.13a*. If the eraser is bent as shown in *figure 7.13b* you can see that on the tension side the lines have moved apart, and on the compression side they have moved together. At the mid-height of the eraser they remain the same distance apart. Obviously where the lines have moved apart a tensile strain is indicated, and where they have moved together a compressive strain is indicated. We could therefore represent the variation in **strain** throughout the depth of the eraser as shown in *figure 7.13c*. Note that the maximum strains occur at the top and bottom faces, and in the middle the strain (and hence the stress) is zero. The level where the strain is zero is known as the **neutral axis**.

If we now make the assumption that the eraser is made from a linear elastic material this implies that stress is proportional to strain. In fact stress = $E \times$ strain, where E is the constant modulus of elasticity. The variation in stress throughout the depth of the eraser must therefore be of the same shape as the variation in strain (*figure 7.13c*).

Let us now switch to an 'I' beam cantilever as shown in *figure 7.14a*, which is assumed to be made from a steel with an elastic perfectly plastic stress/strain relationship as shown in *figure 7.14b*. The aim is to understand what happens to the stresses on the cross-section of the beam at the point of maximum bending moment, *y–y*, as the load, *P*, is gradually increased until ultimate failure occurs.

Initially the load, P_1, is modest so that it only produces strains ε_1 well below the yield strain ε_y. Consequently the accompanying stresses are also well below yield, as shown in *figure 7.14c*. It is interesting to compare the effectiveness of the steel at different levels in contributing to the bending resistance of the beam. Level *a* is close to the neutral axis and level *b* is within the flange of the beam. Remember, the aim is to provide an anticlockwise moment throughout the cross-section in order to cancel the clockwise moment induced by the load *P* acting at distance *L*. Compared to level *a* the steel at level *b* is more effective because:

1. It is subject to higher strains and hence stresses.

2. The extra width of the flange means there is a bigger area of steel available for 'pushing' and 'pulling' (force = stress × area).

3. The moment resisted is obviously proportional to the distance between the forces. Hence the further the steel is from the neutral axis, the more effectively it works.

We shall now increment the load to P_2, which is the load which just causes the strains in the extreme top and bottom fibres of the beam to reach the yield stress – σ_y (*figure 7.14d*). This is the highest load at which the deflection of the beam is fully recoverable. If the load is removed the beam will spring back to its original position. The beam is now at the **limit of elastic bending**.

A further small increment of load to P_3 takes the steel near the top and bottom surfaces beyond its yield point. *Figure 7.14e* shows how, beyond the yield point, as the strain continues to increase the stress remains constant at its yield value. However, closer to the centre of the beam, where the strain is less than σ_y, the steel remains elastic. The beam is not yet at its ultimate failure state, although some permanent deflection has occurred which is not recoverable when the load is removed. This is **elastic-partially plastic bending**.

If the load is incremented further to P_4 the first thing that becomes obvious is the way that relatively small additions to the

Figure 7.14a
Cantilever beam

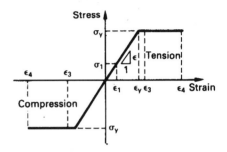

Figure 7.14b
Assumed stress/strain relationship for cantilever material

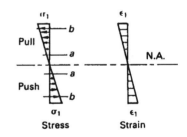

Figure 7.14c
Elastic stresses and strains

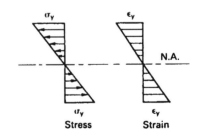

Figure 7.14d
Stresses reach the yield point

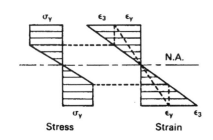

Figure 7.14e
The yield strain exceeded

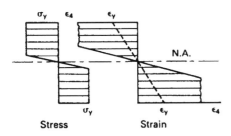

Figure 7.14f
Most of the steel is at the yield stress

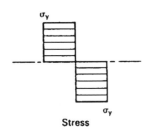

Figure 7.14g
Full yield

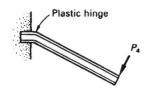

Figure 7.14h
A plastic hinge has formed

load produce large deflections. Because of the ductile nature of the mild steel it can be strained to say fifteen times its yield strain, as shown in *figure 7.14f*, ensuring that virtually all the steel has yielded. As already explained, the material close to the neutral axis makes a very small contribution to bending resistance, so for all practical purposes the small remaining elastic zone on the neutral axis can be ignored. A stress distribution as shown in *figure 7.14g* can thus be assumed.

Once **full plastic bending** has been reached the beam would rotate at constant load (in theory) about the point of maximum bending moment which has become a **plastic hinge** (*figure 7.14h*). Obviously ultimate collapse has been reached. For standard 'I' beams the load to cause ultimate failure, P_4, is about 15% more than the load to cause first yield, P_2.

7.5 Plastic theory of bending

In limit state design it is of prime importance to ensure that a structure has an adequate safety factor against reaching the ultimate limit state or collapse. As plastic bending is taking place at collapse we must derive a numerical relationship between the collapse moment, the yield stress of the material and the cross-sectional dimensions of the beam.

Initially we shall assume that all beams are provided with lateral restraint. This means that they will not buckle sideways before the full yield stress of the material is reached. Beams without lateral restraint are considered later in *section 7.8*.

7.5.1 Rectangular cross-section in plastic bending

Consider a portion of rectangular beam subject to its ultimate bending moment, M_{ult}, so that it is on the point of collapse. By considering the horizontal equilibrium of the beam free body in *figure 7.15* we can see that tensile force = compressive force and hence that the neutral axis must lie at the mid-height of the beam.

$$\text{force} = \text{stress} \times \text{area}$$
$$\text{'pull' force} = \text{'push' force} = \sigma_y \times B \times D/2$$

The distance between the centroids of the two forces or **lever arm** = $D/2$

therefore

$$M_{ult} = \sigma_y \times BD^2/4$$

or

$$M_{ult} = \sigma_y \times S \qquad [7.2]$$

where S is the **plastic section modulus**, and equals $BD^2/4$ for a rectangular section. The plastic section modulus can be seen to be a **geometric** property of the cross-section of the beam. This means that it depends only upon the size and shape of the beam, and is independent of the beam material. It is important, however, because it is a direct measure of the ultimate plastic bending strength of a beam. We shall now look at the way the plastic section modulus, S, varies for differently shaped cross-sections.

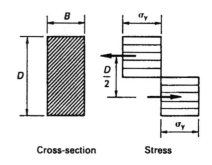

Figure 7.15
Rectangular beam at full yield

7.5.2 "I" beams in plastic bending

Again, for horizontal equilibrium, the total tensile force = total horizontal force and so for a standard 'I' beam with equal flanges the neutral axis must lie at the mid-height of the beam. If we divide the cross-section into four rectangles as shown in *figure 7.16* and take moments about the neutral axis for each rectangle we get:

$$M_{ult} = \sigma_y \times ((A_1 \times Y_1) + (A_2 \times Y_2) + \ldots \text{ etc.})$$

where A_1 etc. is the area of each rectangle, and Y_1 etc. is the distance from the neutral axis to the centroid of each rectangle.

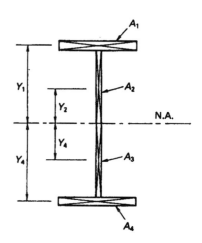

Figure 7.16
Dividing a cross-section into rectangles

However, if the flanges are equal, this simplifies to:

$$M_{ult} = \sigma_y \times 2 \times ((A_1 \times Y_1) + (A_2 \times Y_2)$$

$$= \sigma_y \times \Sigma AY = \sigma_y \times S$$

In this case the plastic section modulus, S, is defined as the **first moment of area** of the section **about the neutral axis**. Values of S are given in section tables for standard sections (see the *Appendix*). Therefore to obtain a suitable beam size for a given ultimate moment we simply use:

$$\text{required } S = \frac{M_{ult}}{\sigma_y} \qquad [7.3]$$

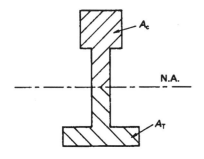

Figure 7.17
A non-standard shape

and select a beam from the section tables with an adequate plastic section modulus, *S*. Always aim for a section whose value is equal to, or slightly greater than, the required value. Normally steel is paid for by weight and so the lightest section that will do the job is usually the most economic choice.

7.5.3 *General cross-sections in plastic bending*

It is now an easy matter to extend these concepts to sections of any cross-sectional shape (*figure 7.17*). To find the ultimate moment that a beam will support it is simply a case of evaluating the plastic section modulus, *S*, and using *equation [7.2]*. For horizontal equilibrium the neutral axis must divide the section into two equal areas:

$$A_{\mathrm{C}} = A_{\mathrm{T}}$$

We then proceed to sum the first moments of area about the neutral axis, as before, in order to obtain the plastic section modulus, *S*.

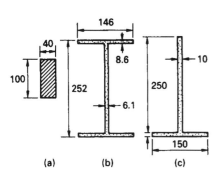

All dimensions in mm

Figure 7.18
Beam cross-sections

Example 7.3

The three beam cross-sections shown in *figure 7.18a–c* all contain the same amount of steel. Find the plastic section modulus, *S*, for each one.

Solution

Comment – The standard section tables (see the Appendix) give the values of S in cm³. Although these are non-standard SI units, they produce more manageable numbers, and it is proposed that we also work in centimetres.

Beam (a)

From *section 7.5.1* $S = BD^2/4 = 4 \times 10^2/4$
Answer   **$S = 100$ cm³**

Beam (b)

From *section 7.5.2* – refer to *figure 7.19a*

All dimensions in cm.

Item	Area, A	y	Ay
1	14.6 × 0.86 = 12.56	12.17	152.9
2	11.74 × 0.61 = 7.16	5.87	42.0
			194.9

$$S = 2 \times 194.9$$

Answer $$S = 389.8 \text{ cm}^3$$

Comment – The above beam is a standard 254 × 146 × 31 kg/m universal beam, and consequently we can check this result from the standard section table given in the Appendix. The value given is 396 cm³. The small difference is due to rounding error and the presence of fillet radii on the actual section.

Beam (c) – refer to figure 7.19b

The first step is to calculate the depth of the neutral axis, Y, which must divide the area into two equal halves. After this step the table can be completed.

$$\text{Total area} = (1 \times 25) + (1 \times 15) = 40 \text{ cm}^2$$

$$\text{Depth, } Y = \frac{40}{2 \times 1} = 20 \text{ cm}$$

All dimensions in cm.

Item	Area, A	y	Ay
1	1 × 20 = 20	10	200
2	1 × 5 = 5	2.5	12.5
3	1 × 15 = 15	5.5	82.5
			295.0

Answer $$S = 295 \text{ cm}^3$$

Comment – The above results show that, as expected, the 'I' beam has the biggest plastic section modulus, and hence is the most efficient for resisting bending.

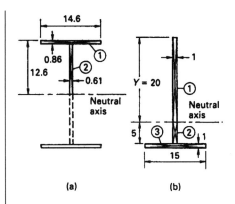

Figure 7.19
Divide the sections into rectangles to evaluate the first moments of area

Example 7.4

In *example 7.2* we used an approximate method to calculate the maximum ultimate load that can be placed on the end of a 3 m long cantilever made from a $254 \times 146 \times 31$ kg/m universal beam. Repeat the example assuming full plastic bending of the whole section.

Solution

From section tables (see *Appendix*) the plastic modulus, S, for a $254 \times 146 \times 31$ kg/m universal beam = 396 cm³. As before, the yield stress, $\sigma_y = 275$ N/mm².

From *equation [7.3]*

$$S = \frac{M_{ult}}{\sigma_y}$$

therefore

$$M_{ult} = S \times \sigma_y = 396 \times 275 \times 10^{-3}$$
$$= 108.9 \text{ kNm}$$

As in *example 7.2*

maximum $\qquad M = 3 \times P$

hence at failure $\qquad 3 \times P = 108.9$ kNm

Answer $\qquad\qquad \boldsymbol{P = 36.3 \text{ kN}}$

Comment – This is a 30% increase compared to the value of 28.0 kN obtained in example 7.2. This represents the contribution of the web to the bending strength.

7.6 Shear stresses in beams

If you take a flexible plastic ruler and apply a shear force to it (*figure 7.20*), you will notice how it is very much stiffer when the loads are applied parallel to the flat surface of the ruler, compared to when the ruler is laid flat. This suggests that if a shear force is applied to an 'I' beam, most of it will be resisted by the web. This is indeed the case. In fact it is usual to assume that **only the web resists shear**.

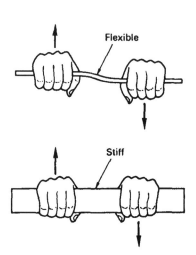

Figure 7.20
Shear forces applied to a flexible ruler

Consider a portion of beam subjected to a shear force (*figure 7.21a*). If we consider a small element of the web (*figure 7.21b*) the shear forces will produce shear stresses, τ, on the vertical sides. However, the element is not in moment equilibrium. Equal shear stresses must be present on the horizontal faces of the element to prevent it spinning in a clockwise direction. These are known as **complementary shear stresses** (*figure 7.21c*). These stresses can be resolved as shown in *figure 7.21d*. Thus shear forces can be considered to produce diagonal tensile and compressive forces in the web of a beam. In a thin ductile material, such as the web of a steel beam, the compressive force can cause web buckling (*figure 7.22a*). In a brittle material, such as concrete, the tensile force can cause tensile cracking (*figure 7.22b*). Inspection of the shear force diagrams that we have drawn so far indicates that maximum shear forces occur at supports. These are therefore usually the critical points.

The magnitude of the shear stress, τ, is reasonably constant throughout the depth of the web. It is therefore only necessary to check that the **average** value of the shear stress is less than a permitted ultimate value – usually taken as $0.6 \times$ yield stress, σ_y.

$$\text{Average shear stress on web} = \frac{\text{shear force}}{\text{web area}} = \frac{V}{D \times t}$$

D is the total depth of the beam and t is the thickness of the web.

For mild steel this should be less than $0.6 \times 275 = 165$ N/mm^2

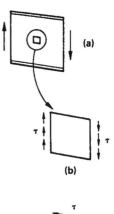

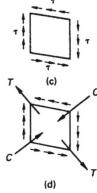

Figure 7.21
Shear stresses on the web of a beam

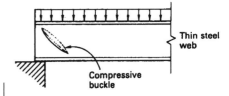

Figure 7.22a
Shear failure in a steel web

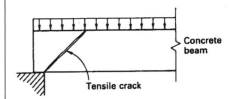

Figure 7.22b
Shear failure in a concrete beam

Example 7.5

In *example 4.8* we calculated the loading on two beams which form the support structure for the roof of a double garage. Determine the dimensions of suitable standard universal beams.

Solution

Figure 7.23 shows the loading on the two beams from *example 4.8*.
The first stage is to evaluate the beam reactions and then to draw the shear force and bending moment diagrams.

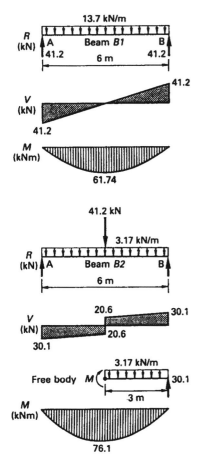

Figure 7.23
The beams over a double garage

Beam B1

From symmetry:

$$\text{Reactions}\quad R_A = R_B = 13.72 \times 6/2 = 41.2 \text{ kN}$$

The resulting diagrams are shown in *figure 7.23*. Beam B1 is a standard case – i.e. a simply supported beam with a uniformly distributed load. We saw in *section 7.3.4* that for this case:

$$M_{max} = \frac{wL^2}{8} = \frac{13.72 \times 6^2}{8}$$

$$= 61.74 \text{ kNm}$$

From *table 3.4* a typical yield stress for mild steel is 275 N/mm^2.

From *equation [7.3]*

$$\text{Required}\quad S = \frac{M_{ult}}{\sigma_y} = \frac{61.74 \times 10^6}{275 \times 10^3}$$

$$= 224.5 \text{ cm}^3$$

From the *Appendix*
Try 254 × 102 × 22 kg/m universal beam
(S = 262 cm³, D = 254 mm, t = 5.8 mm)

$$\text{Average shear stress} = \frac{41.2 \times 10^3}{254 \times 5.8} = 28 \text{ N/mm}^2$$

$$(< 165 \text{ N/mm}^2)$$

Answer   Use 254 × 102 × 22 kg/m universal beam

Beam B2

From symmetry:

$$\text{Reactions}\quad R_A = R_B = \frac{(3.17 \times 6) + 41.2}{2}$$

$$= 30.1 \text{ kN}$$

The shear force diagram is quite straightforward provided you obey the guidelines – start at the right and draw in the direction of the force (*figure 7.23*). The peak bending moment again occurs at mid-span, and can be obtained in either of two ways. Firstly consider the free body diagram shown:

$$M = (30.1 \times 3) - (3.17 \times 3 \times 3/2)$$
$$= 76.0 \text{ kNm}$$

The second method is to combine two standard cases. We have a simply supported beam with a point load at mid-span ($M_{max} = PL/4$), plus a simply supported beam with a UDL throughout ($M_{max} = wL^2/8$). As the maximum bending moments both occur at mid-span we can simply add the two cases together. This is an example of what is known as **the principle of superposition**.

$$M = \frac{41.2 \times 6}{4} + \frac{3.17 \times 6^2}{8}$$

$$= 61.8 + 14.3 = 76.1 \text{ kNm}$$

required
$$S = \frac{M_{ult}}{\sigma_y} = \frac{76.1 \times 10^6}{275 \times 10^3}$$

$$= 276.7 \text{ cm}^3$$

From the *Appendix*
Try 254 × 102 × 25 kg/m universal beam
(S = 306 cm³, D = 257 mm, t = 6.1 mm)

$$\text{Average shear stress} = \frac{30.1 \times 10^3}{257 \times 6.1} = 19.2 \text{ N/mm}^2$$

($< 165 \text{ N/mm}^2$)

Answer   Use 254 × 102 × 25 kg/m universal beam

Comments:

1. It can be seen from the above that the average shear stresses are well below the permitted values. This is typical for most beams and consequently many designers do not bother to check shear. It can, however, be a problem if a large point load occurs close to a support. This causes large shear forces without necessarily increasing the bending moment very much.

2. The above two beams were selected from section tables (see the Appendix) on the basis that they are the lightest sections to have an adequate plastic section modulus. In practice, if the steel beams have to be ordered from a steel stockholder, it may be more economical to order two beams of the same size and weight. This means that they would have to be the heavier section.

7.7 Elastic theory of bending

Limit state design puts great stress on securing an adequate safety factor against ultimate collapse, which means considering fully plastic failure in most cases. There are, however, situations where it is necessary to carry out an elastic analysis, and these include:

1. In the design of structures subject to frequent variations in stress, such as cranes, the stresses must be kept well below the yield value in order to provide adequate protection against fatigue failure. Only an elastic analysis can give an indication of the actual magnitude of stresses under normal working loads.

2. Most mechanical engineering structures are not covered by national standards and the traditional design methods employ an elastic approach.

3. Steel beams with relatively thin components may buckle locally before the fully plastic moment is reached, and their performance must therefore be restricted to the elastic range. Only very few standard beam sections are affected in this way.

4. Timber members are often still designed using elastic principles.

5. Pre-stressed concrete designs are often governed by serviceability criteria, such as keeping tensile stresses below a certain value, rather than the ultimate collapse criterion. An elastic approach must be adopted to make these serviceability checks.

7.7.1 Rectangular cross-section in elastic bending

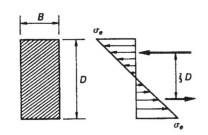

Figure 7.24
A rectangular beam with elastic bending

Consider a portion of rectangular beam subject to a bending moment, M, which causes a maximum stress, σ_{max}, which is less than the yield stress, σ_y (*figure 7.24*). We must still satisfy horizontal equilibrium and hence the **total tensile force = total compressive force**. For symmetrical sections, therefore, the neutral axis must lie at mid-height as in plastic analysis. We need to find the relationship between the stress, σ_{max}, the bending moment, M, and the beam cross-sectional dimensions, B and D.

pull force = push force = mean stress × area

$$= \sigma_{max}/2 \times B \times D/2$$
$$= \sigma_{max} \times BD/4$$

For a triangle the centroid is one-third of the height from the base, therefore the distance between the centroids of the two forces:

lever arm = $2/3 \times D$

therefore $M = \sigma_{max} \times BD/4 \times 2D/3$
$$= \sigma_{max} \times BD^2/6$$

or $M = \sigma_{max} \times Z$ [7.4]

where Z is the **elastic section modulus**, and equals $BD^2/6$ for a rectangular section. The elastic section modulus, Z, is obviously analogous to the plastic section modulus, S. (Take care not to confuse the 'elastic section modulus' with the 'modulus of elasticity'. There is no connection. The first is a geometric property, based on the shape and dimensions of a section. The second is a material stiffness property.)

7.7.2 *General cross-sections in elastic bending*

Because the stress varies throughout the section, the above approach would get too tedious for more complex shapes and we need to develop a more elegant and general theory.

Consider the beam cross-section shown in *figure 7.25a*, which has a value of stress of σ_{max} at the distance Y from the neutral axis. The first step is to determine the position of the neutral axis. We must still satisfy horizontal equilibrium on the section and thus ensure that tensile forces = compressive forces.

Consider a small element of area, a, shown in *figure 7.25a*. We know, from similar triangles, that the stress on this element is proportional to the distance of the element, y, from the neutral axis:

$$\text{stress} = \frac{\sigma_{max} \times y}{Y}$$

and   force = stress × area

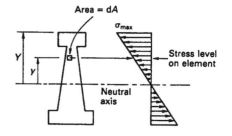

Figure 7.25a
A non-standard beam

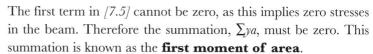

For the whole cross-section:

$$\Sigma(\sigma_{max}/Y \times y.a) = \frac{\sigma_{max}}{Y} \times \Sigma(y.a)$$

If y is considered positive above the neutral axis and negative below, then for horizontal equilibrium (tensile forces = compressive forces) the sum of forces over the whole cross-section must be zero.

$$0 = \frac{\sigma_{max}}{Y} \times \Sigma(y.a) \qquad [7.5]$$

The first term in *[7.5]* cannot be zero, as this implies zero stresses in the beam. Therefore the summation, Σya, must be zero. This summation is known as the **first moment of area**.

Now consider a thin uniformly thick slice of the cross-section balanced over a knife edge support as shown in *figure 7.25b*. Clearly for equilibrium the centre of gravity or **centroid**, C, of the section must be placed directly over the support. As with a seesaw, if equilibrium is to be maintained, the net sum of each element of area, a, multiplied by its distance to the pivot, y, must be zero. Writing this in mathematical terms we get:

$$\Sigma y.a = 0 \qquad [7.6]$$

By comparing *equations [7.5]* and *[7.6]* above we can conclude that the **neutral axis passes through the centroid**. (Notice that this is not the same as for plastic bending and hence the neutral axis must move as the bending changes from elastic to plastic.)

Now take moments about the neutral axis for each element of force. Remember moment = force × distance. This means that each element within the summation must be multiplied by y again.

$$M = \frac{\sigma_{max}}{Y} \times \Sigma(y^2.a)$$

The summation $\Sigma(y^2.a)$ is termed the **second moment of area**, **I**. (It is similar to, and often erroneously called, the moment of inertia, but strictly speaking moment of inertia should contain a mass term in place of the area.)

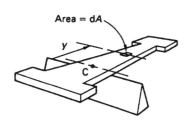

Figure 7.25b
The beam shape balances when the centroid is over the knife edge

We therefore get

$$M = \frac{\sigma_{max}}{Y} \times I$$

or

$$\frac{M}{I} = \frac{\sigma_{max}}{Y} \qquad [7.7]$$

Values of I for standard beams are given in section tables (see the *Appendix*), and we shall look at how to evaluate I for non-standard shapes of cross-section later.

   Equation [7.7] is very important and is sometimes known as the **engineer's equation of bending**. It can be used in many ways:

Question 1 – For a given beam what is the maximum moment that can be supported if the stress is not to exceed a given maximum value of σ_{max}?

Rearranging *[7.7]*

$$M = \sigma_{max} \times \frac{I}{Y}$$

but from *[7.4]*

$$M = \sigma_{max} \times Z$$

hence **elastic section modulus**

$$Z = \frac{I}{Y}$$

Thus Z is a geometric property of a beam which is directly proportional to its bending strength. Values of Z are also given in section tables (see the *Appendix*) for standard beams. We shall look at how to calculate Z values for non-standard beams later.

Example 7.6

Figure 7.26 shows a house floor which uses 50 mm wide × 150 mm deep timber joists spaced at 400 mm centres. If the maximum permissible stress in the timber is limited to 7.5 N/mm², determine the maximum allowable simply supported span of the floor if it supports the following loads:

$$\begin{aligned} \text{dead load of floor} &= 0.45 \text{ kN/m}^2 \\ \text{imposed load on floor} &= 1.5 \text{ kN/m}^2 \end{aligned}$$

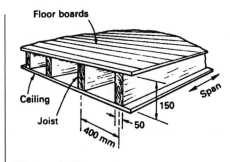

Figure 7.26
Timber floor construction

Solution

Comment – It has already been stated that timber structures are often still designed using the permissible stress philosophy (section 1.7.1). In this approach partial safety factors are not applied to the loads.

$$\text{Total load} = 0.45 + 1.5 = 1.95 \text{ kN/m}^2$$
$$\text{Load/m, } w, \text{ on each joist} = 1.95 \times 0.4 = 0.78 \text{ kN/m}$$

For a simply supported floor joist with a uniformly distributed load:

$$M = \frac{wL^2}{8} = \frac{0.78L^2}{8}$$

For a beam with a rectangular cross-section:

$$\text{Elastic section mod. } Z = \frac{BD^2}{6} = \frac{50 \times 150^2}{6}$$

$$= 187\ 500 \text{ mm}^3$$

From above $M = \sigma_{max} \times Z = 7.5 \times 187\ 500$
$$= 1\ 406\ 000 \text{ Nmm}$$
$$= 1.406 \text{ kNm}$$

Equating the two values of M:

$$1.406 = \frac{0.78L^2}{8}$$

From this $L = 3.79$ m

Answer   Allowable span = 3.79 m

Question 2 – Given a moment and a maximum allowable stress, what beam should be used?

Rearranging [7.14] $$\frac{I}{Y} = \frac{M}{\sigma_{max}}$$

and $$Z = \frac{I}{Y}$$

$$Z = \frac{M}{\sigma_{max}}$$

Therefore use a beam with a Z value at least equal to that required by the above formula. In the case of standard beams, look in section tables for a beam with a suitable Z value.

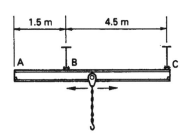

Figure 7.27a
A crane hoist

Example 7.7

Figure 7.27a shows a crane hoist which slides on the bottom flange of a universal beam, ABC. It is required to have a safe working load of 30 kN which should be increased by 25% to allow for sudden impact loading. The hoist and chain etc. weigh 1 kN. The self-weight of the beam can be ignored. If the following stresses are not to be exceeded, determine a suitable standard universal beam:

bending tension	165 N/mm^2
bending compression	60 N/mm^2

Solution

The load to be used in calculations is the safe working load increased by 25% plus the weight of the hoist and chain:

$$\text{Load} = (30 \times 1.25) + 1 = 38.5 \text{ kN}$$

This is a point load which can act anywhere along the beam. We need to find the position which leads to the maximum possible bending moment. Two cases need to be checked.

Case 1 – load mid-way between B and C

see *figure 7.27b* $$M = \frac{P \times L}{4} = \frac{38.5 \times 4.5}{4}$$

$$= 43.3 \text{ kNm}$$

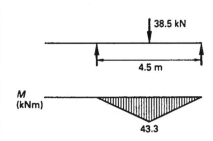

Figure 7.27b
Load case 1

Case 2 – load at A

see *figure 7.27c* $M = P \times 1.5 = 38.5 \times 1.5$
$$= 57.75 \text{ kNm}$$

Therefore *case 2* is critical. For a standard 'I' beam the stresses in tension and compression will be equal. The allowable stress must therefore be limited to the compressive value of 60 kN/mm^2.

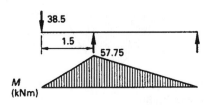

Figure 7.27c
Load case 2

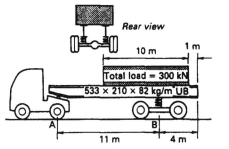

Figure 7.28a
A load on the trailer of a heavy goods vehicle

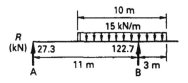

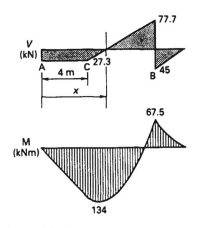

Figure 7.28b
Shear force and bending moment diagrams

From
$$Z = \frac{M}{\sigma_{max}} = \frac{57.75 \times 10^6}{60 \times 10^3}$$

$$= 962.5 \text{ cm}^3$$

Answer   Use 406 × 178 × 60 kg/m universal beam
$$(Z = 1060 \text{ cm}^3)$$

Question 3 – Given a moment and a particular beam, what is the maximum stress in the beam?

Rearranging *[7.7]*
$$\sigma_{max} = \frac{M \times Y}{I}$$

$$= \frac{M}{Z}$$

Example 7.8

Figure 7.28a shows a heavy-goods vehicle trailer with the main chassis members consisting of two $533 \times 210 \times 82$ kg/m universal beams, which can be considered simply supported at A and B. If the vehicle carries a load of 300 kN in the position shown, what is the maximum bending stress in the beam. Assume the load is uniformly distributed and ignore the self-weight of the beams. Also determine the average shear stress on the web.

Solution

If the load is assumed to be evenly divided between the two beams:

$$\text{Distributed load/beam} = 150/10 = 15 \text{ kN/m}$$

$(\Sigma M \text{ about B})$
$$\text{reaction, } R_A \times 11 = 15 \times 10 \times 2$$
$$R_A = 27.3 \text{ kN}$$

Comment – The '2' in the above equation is the distance from B to the centre of the uniformly distributed load.

$(\Sigma V = 0)$ $R_B = 150 - 27.3 = 122.7$ kN

The shear force diagram is shown in *figure 7.28b*.

A peak bending moment occurs at the point of zero shear which is distance x from A:

$$\text{Distance } x = 4 + 27.3/15 = 5.82 \text{ m}$$
$$M_{\text{Peak}} = (27.3 \times 5.82) - (15 \times 1.82 \times 0.91)$$
$$= 134 \text{ kNm}$$

Also
$$M_{\text{B}} = 15 \times 3 \times 1.5$$
$$= 67.5 \text{ kNm}$$

and
$$M_{\text{C}} = 27.3 \times 4$$
$$= 109 \text{ kNm}$$

The bending moment diagram is also shown in *figure 7.28b*. From section tables, the elastic section modulus, Z, for a $533 \times 210 \times 82$ kg/m universal beam is 1800 cm^3 and $D = 528.3$ mm, $t = 9.6$ mm.

Hence
$$\sigma_{\text{max}} = \frac{M}{Z} = \frac{134 \times 10^6}{1800 \times 10^3}$$

Answer   max. bending stress = 74.4 N/mm^2

From the shear force diagram:

$$V_{\text{max}} = 77.7 \text{ kN}$$

$$\text{Average shear stress} = \frac{\text{shear force}}{\text{web area}} = \frac{77.7 \times 10^3}{528.3 \times 9.6}$$

Answer   Av. shear stress = 15.3 N/mm^2

Question 4 – Given a moment and a particular beam, what is the stress at a certain distance, y, from the neutral axis?

We saw above, from similar triangles, that the stress at any distance, y, from the neutral axis is given by:

$$\text{stress, } \sigma = \frac{\sigma_{\text{max}} \times y}{Y}$$

therefore
$$\frac{\sigma}{y} = \frac{\sigma_{\text{max}}}{Y}$$

Thus our original engineer's equation can be rewritten:

$$\frac{M}{I} = \frac{\sigma}{y} \qquad\qquad [7.8]$$

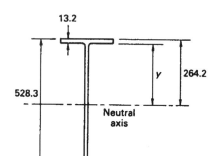

Figure 7.29
The dimensions of a $533 \times 210 \times 82$ kg/m universal beam. All dimensions in mm

This is a more fundamental form as it can be used to evaluate the stress at any point in a beam, not just at the top and bottom.

Rearranging *[7.8]* $$\sigma = \frac{M \times y}{I}$$

Example 7.9

For the chassis beams in *example 7.8*, what is the bending stress at the junction between the web and the flange?

Solution

From section tables, the dimensions for a $533 \times 210 \times 82$ kg/m universal beam are as shown in *figure 7.29*. Also from section tables, the second moment of area, I, is 47 500 cm^4.

$$\text{Distance, } y = 264.2 - 13.2 = 251 \text{ mm}$$

hence $$\sigma = \frac{M \times y}{I} = \frac{134 \times 106 \times 251}{47\ 500 \times 10^4}$$

Answer   bending stress = 70.8 N/mm^2

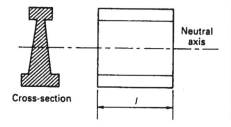

Figure 7.30a
A short length of beam before bending

Before we leave the engineer's theory of bending it is necessary to link bending moment and stress with strain. This will be used later when we come to consider the buckling of compression members and the deflection of beams. If you are not interested in the derivation you can skip the following section.

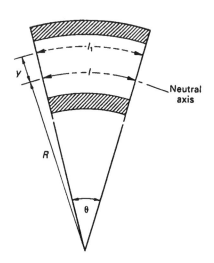

Figure 7.30b
A short length of beam after bending

Consider the short length of beam shown unloaded in *figure 7.30a*. If a bending moment is applied, the beam is assumed to take the curved shape shown in *figure 7.30b*. It has a radius of R. We know that there will be no strain, and hence stress, on the neutral axis. Therefore the length at this level will remain l. However, above the neutral axis the curvature will cause the material to be in tension and extend, and below the neutral axis it will be in compression and contract. We can say that:

$$\text{length, } l = R\theta \quad (\theta \text{ in radians})$$

At an arbitrary level, distance y from the neutral axis:

$$\text{length, } l_l = (R + y)\theta$$

Strain at this level $= \dfrac{\text{extension}}{\text{original length}} = \dfrac{l_1 - l}{l}$

$$= \frac{(R + y)\theta - R\theta}{R\theta} = \frac{y}{R}$$

Stress at this level, $\sigma = E \times \text{strain} = \dfrac{Ey}{R}$

or $\qquad \dfrac{\sigma}{y} = \dfrac{E}{R}$

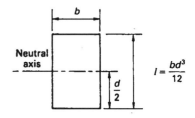

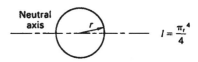

Figure 7.31

Second moments of area for simple geometric shapes

Thus the engineer's equation of bending can be extended to:

$$\frac{M}{I} = \frac{\sigma}{y} = \frac{E}{R} \qquad [7.9]$$

7.7.3 Second moments of area for simple shapes

We have seen how I values for standard sections are given in tables. Values of I for simple geometric shapes can be obtained by integration and are widely published. Two are shown in *figure 7.31*.

Holes in sections can be treated as negative areas. For example the two sections shown in *figure 7.32* have a second moment of area given by:

$$I_{\text{NA}} = \frac{bd^3}{12} - \frac{\pi r^4}{4}$$

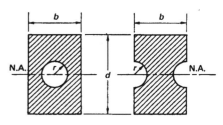

Figure 7.32

Simple shapes with holes

7.7.4 Section properties for non-standard beams

This section is concerned with the calculation of the second moment of area, I, and the elastic section modulus, Z, for non-standard shapes. We shall also derive an expression for a third geometric property known as the **radius of gyration**, which will be used in the next chapter for the design of compression members. We shall concentrate at first on finding the second moment of area.

Consider the cross-section shown in *figure 7.33*. Because this is an unsymmetrical beam (i.e. the top and bottom flanges are unequal) we first need to find the position of the neutral axis. We have already proved that it passes through the centroid, C, of the section. The process starts by dividing the cross-section into items

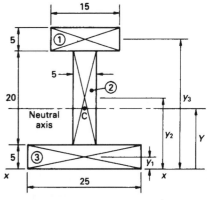

All dimensions in cm.

Figure 7.33

A non-standard beam

of simple shape. In this case that consists of the three rectangles numbered 1–3. The distance, Y, from any arbitrary axis – say x–x, to the neutral axis is given by:

$$Y = \frac{\Sigma Ay_n}{\Sigma A} \qquad [7.10]$$

where: A = area of each item

y_n = distance from the arbitrary x–x axis to the centroid of each item

ΣAy_n = sum of the first moments of area.

The whole calculation is best performed in tabular form for all but the simplest of cases.

All dimensions are in cm

Item A	Area	y_n	Ay_n
1	75	27.5	2063
2	100	15	1500
3	125	2.5	312
	300		3875

From this $Y = \dfrac{\Sigma Ay_n}{\Sigma A} = \dfrac{3875}{300}$

$$= 12.92 \text{ cm}$$

The next step is to use the **parallel axes theorem** to determine the second moment of area, I, of each item about the neutral axis. The parallel axes theorem is used to evaluate the I of a particular item about any axis parallel to its own. If the distance from the neutral axis of the item to the new axis is y, the theorem states:

$$I_{NA} = I_{self} + Ay^2 \qquad [7.11]$$

where: I_{self} = second moment of area of the element about its own neutral axis – from *figure 7.31*

A = area of the item

y = distance from the neutral axis of the item to the new axis

or $I_{NA} = I_{self} + I_{transfer}$

$$\text{total } I_{NA} = \Sigma I_{self} + \Sigma I_{transfer} \qquad [7.12]$$

The previous tabular calculations will now be extended:

Item	Area A	yn	Ayn	y	$I_{transf.}$ Ay^2	I_{self}
1	75	27.5	2063	14.58	15 943	$15 \times 5^3/12 =$ 156
2	100	15.0	1500	2.08	433	$5 \times 20^3/12 = 3333$
3	125	2.5	312	10.42	13 572	$25 \times 5^3/12 =$ 260
	300		3875		29 948	3749

From *equation [7.12]*

$$I_{NA} = \Sigma I_{self} + \Sigma I_{transfer}$$
$$= 3749 + 29\ 948 = 33\ 697 \text{ cm}^4$$

We can now proceed to determine the elastic section moduli. Because the neutral axis does not lie at the mid-height of the beam, there will be two elastic section moduli to consider – Z_{top} and Z_{bottom}. This of course implies that, when subjected to a bending moment, the maximum tensile and compressive stresses will not be equal.

Elastic section moduli:

$$Z_{top} = \frac{I}{Y_{top}} = \frac{33\ 697}{(30 - 12.92)} = 1973 \text{ cm}^3$$

$$Z_{bottom} = \frac{I}{Y_{bottom}} = \frac{33\ 697}{12.92} = 2608 \text{ cm}^3$$

hence maximum stresses at top and bottom of beam are given by:

$$\sigma_{top} = \frac{M}{Z_{top}}$$

$$\sigma_{bottom} = \frac{M}{Z_{bottom}}$$

The radius of gyration, r, of a section is defined as the radius at which the whole area can be considered to be concentrated in order to have the same second moment of area. The self-inertia of such a small area is zero. The inertia of such an area about the neutral axis is therefore given by the transfer inertia only. From the parallel axes theorem:

$$I_{NA} = Ar^2$$

hence

$$\textbf{radius of gyration, } r = \sqrt{(I_{NA}/A)} \qquad [7.13]$$

We shall use the radius of gyration later, when we come to consider the design of unrestrained beams and columns.

Example 7.10

A $254 \times 146 \times 31$ kg/m universal beam has a 175 mm $\times 10$ mm plate welded to the top flange. If fatigue considerations limit the maximum permissible stresses to the following values, determine the maximum sagging and hogging bending moments to which the beam can be subjected.

$$\text{Maximum compressive stress} = 120 \text{ N/mm}^2$$
$$\text{Maximum tensile stress} = 165 \text{ N/mm}^2$$

Solution

Comment – Because values for I and A are available in section tables for the standard beam, we can consider it as one 'item', rather than dividing it into three rectangles.

From the *Appendix*, for a $254 \times 146 \times 31$ kg/m universal beam, $A = 40.0$ cm^2 and $I = 4\,439$ cm^3. Refer to *figure 7.34*.

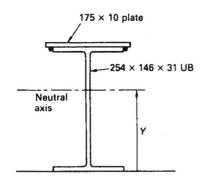

Figure 7.34
A standard beam with a welded top-plate

Item	Area A	y_n	Ay_n	y	$I_{transf.}$ Ay^2	I_{self}
plate	17.5	25.65	449	9.09	1446	$17.5 \times 1^3/12 = \quad 1$
UB	40.0	12.58	503	3.98	634	4439
	57.5		952		2080	4440

$$\text{from this} \qquad Y = \frac{\Sigma Ay_n}{\Sigma A} = \frac{952}{57.5}$$

$$= 16.56 \text{ cm}$$

$$I_{NA} = \Sigma I_{self} + \Sigma I_{transfer} = 4440 + 2080$$
$$= 6520 \text{ cm}^4$$

Elastic section modului:

$$z_{top} = \frac{I}{r_{top}} = \frac{6520}{(26.15 - 16.56)}$$

$$= 680 \text{ cm}^3$$

$$Z_{\text{bottom}} = \frac{I}{r_{\text{bottom}}} = \frac{6520}{16.56} = 394 \text{ cm}^3$$

Sagging – this produces compression in the top and tension in the bottom. Therefore the maximum permitted sagging bending moment is given by the **lesser** of:

$$\sigma_{\text{comp.}} \times Z_{\text{top}} = 120 \times 680 \times 10^{-3}$$
$$= 81.6 \text{ kNm}$$

or

$$\sigma_{\text{tens.}} \times Z_{\text{bot.}} = 165 \times 394 \times 10^{-3}$$
$$= 65.0 \text{ kNm}$$

Answer $\mathbf{M_{\text{sagging}} = 65.0 \text{ kNm}}$

Hogging – this produces tension in the top and compression in the bottom. Therefore the maximum permitted hogging bending moment is given by the **lesser** of:

$$\sigma_{\text{tens.}} \times Z_{\text{top}} = 165 \times 680 \times 10^{-3}$$
$$= 122.0 \text{ kNm}$$

or

$$\sigma_{\text{comp.}} \times Z_{\text{bot.}} = 120 \times 394 \times 10^{-3}$$
$$= 47.3 \text{ kNm}$$

Answer $\mathbf{M_{\text{hogging}} = 47.3 \text{ kNm}}$

7.8 Beams without lateral restraint

We saw in section *7.5.2* that standard beams could be designed using *equation [7.3]*:

$$\text{required } S = \frac{M_{\text{ult}}}{\sigma_{\text{y}}}$$

This implies that the beam will reach full plastic yield at failure. This is a reasonable assumption for hollow sections. However, for ordinary 'I' beams, this is only true if the top flange of the beam is restrained from sideways movement. Such restraint is usually provided if a floor slab sits directly on the top flange. An unrestrained beam will fail by twisting before the full yield stress is

Figure 7.35

A section of beam subject to lateral torsional buckling

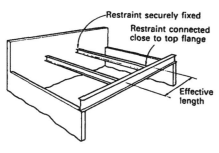

Figure 7.36

Restraints to reduce the effective length of a beam

reached, as shown in *figure 7.35*. This failure mode is known as **lateral torsional buckling**. The theory is beyond the scope of this book, however, the design procedure for unrestrained beams is relatively simple. *Equation 7.3* above is simply modified to:

$$\text{required } S = \frac{M_{\text{ult}}}{p_{\text{b}}} \qquad [7.14]$$

where p_{b} is a **reduced** bending stress to take account of lateral torsional buckling. Values of p_{b} are given in tables or charts and depend upon two factors:

1. The **slenderness ratio** $= \dfrac{L_{\text{E}}}{r_{\text{min}}}$ $\qquad [7.15]$

where L_{E} = effective length
r_{min} = minimum radius of gyration

2. The **torsional index**, *x*. This is given in section tables but can also be taken as equal to the beam depth divided by the flange thickness.

The effective length is obtained by multiplying the actual length by a factor which depends upon the degree of restraint at the ends of the beam. This is often a situation where 'engineering judgement' is called for. The factor can range from 0.5 to 7.5, but for simple beam and column construction the effective length is often taken to be the actual distance between column centres. If intermediate restraints are provided to the top flange, as shown in *figure 7.36*, then the effective length is taken as the distance between restraints.

Figure 7.37 shows the values of the stress, p_{b}, to be used for the design of unrestrained beams in Grade 43 mild steel. The design of unrestrained beams, to support a particular bending moment, becomes a case of trial-and-error as shown in the following example.

Example 7.11

A two-storey car park is to be formed in an existing building (*figure 7.38a*). The elevated floor consists of a 200 mm thick reinforced concrete slab, measuring 12 m × 8 m, which is supported by two parallel steel beams – beams A. It is also supported by the perimeter walls which run parallel to the beams. For ease of access, a

12 m wide opening is formed on the ground floor by using beam B to support the ends of these two beams. Determine the size of suitable standard beams for both beams A and B.

Solution

Loads – From *table 4.1* the unit weight of concrete is 24 kN/m³ and from *table 4.3* an appropriate imposed load for a car park is 2.5 kN/m². Using the normal partial safety factors for loads of 1.4 and 1.6 on dead and imposed loads respectively:

$$\text{Design load of slab} = (1.4 \times 0.2 \times 24) + (1.6 \times 2.5)$$
$$= 10.72 \text{ kN/m}^2$$

Each beam A supports a width of slab of 4 m and is 8 m long.

$$\text{Load on beam A} = 10.72 \times 4 = 42.88 \text{ kN/m}$$
$$\text{say} \qquad\qquad = 45 \text{ kN/m including self-weight}$$

The reaction forces from each beam A will form a point load on beam B.

$$\text{Point loads on beam B} = 45 \times 8/2 = 180 \text{ kN each}$$
$$\text{say} \qquad\qquad\qquad = 185 \text{ kN including self-weight}$$

Comment – The self-weight of beam B is, of course, a uniformly distributed load. However, it is small compared to the point loads. Therefore the above procedure of simply increasing the point loads is acceptable.

Analysis

$$M_{\max} \text{ for beam A} = \frac{wL^2}{8} = \frac{45 \times 8^2}{8} = 360 \text{ kNm}$$

$$M_{\max} \text{ for beam B} = 185 \times 4 = 740 \text{ kNm}$$

The shear force and bending moment diagrams are shown in *figure 7.38b.*

Design

Beams A

The slab will provide full lateral restraint and we can use *equation [7.3]* with $\sigma_y = 275$ for Grade 43 steel:

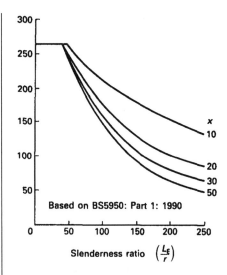

Figure 7.37
Bending stresses for unrestrained beams in Grade 43 steel

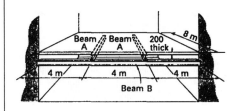

Figure 7.38a
A two-storey car park structure

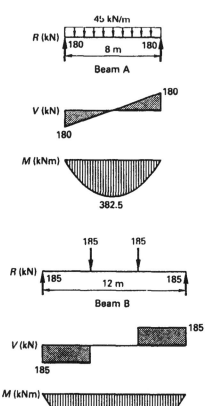

Figure 7.38b
Shear force and bending moment diagrams for beams A and B

$$\text{required } S = \frac{M_{\text{ult}}}{\sigma_y} = \frac{360 \times 10^6}{275 \times 10^3}$$

$$= 1309 \text{ cm}^3$$

Answer   Use 457 × 191 × 67 kg/m universal beam
$$(S = 1470 \text{ cm}^3)$$

Beam B

This beam must be considered unrestrained between the connection points to the other beams.

$$\text{Effective length, } L_{\text{E}} = 4000 \text{ mm}$$

In order to proceed we must now estimate the bending stress, p_{b}, so that a first trial section can be selected. Try $p_{\text{b}} = 150 \text{ N/mm}^2$

From *equation [7.14]*

$$\text{required } S = \frac{M_{\text{ult}}}{p_{\text{b}}} = \frac{740 \times 10^6}{150 \times 10^3}$$

$$= 4933 \text{ cm}^3$$

Try 762 × 267 × 147 kg/m universal beam
($S = 5170 \text{ cm}^3$, $r_{min} = 5.39$ cm, $x = 45.1$).

From *equation [7.15]*

$$\text{Slenderness ratio} = \frac{L_{\text{E}}}{r_{\text{min}}} = \frac{4000}{53.9}$$

$$= 74.2$$

From *figure 7.37* $p_{\text{b}} = 195 \text{ N/mm}^2$

$$\text{Actual stress} = \frac{740 \times 10^6}{5170 \times 10^3} = 143 \text{ N/mm}^2$$

$$< 195 \text{ N/mm}^2$$

Comment – As the actual stress is less than p_b the beam would be safe, however, it is possibly worth trying to improve economy by trying a smaller beam.

Try 686 × 254 × 125 universal beam
($S = 4000 \text{ cm}^3$, $r_{min} = 5.24$ cm, $x = 43.9$).

From *equation [7.14]*

$$\text{Slenderness ratio} = \frac{L_E}{r_{min}} = \frac{4000}{52.4}$$

$$= 76.3$$

From *figure 7.36* $p_b = 196 \text{ N/mm}^2$

$$\text{Actual stress} = \frac{740 \times 10^6}{4000 \times 10^3} = 185 \text{ N/mm}^2$$

$$< 196 \text{ N/mm}^2$$

Answer   Use 686 × 254 × 125 kg/m universal beam

Comment – Strictly speaking we should check the value of the average shear stress on the webs. However, as already stated, this is rarely a problem unless we have large point loads close to supports.

7.9 Rectangular reinforced concrete beams (BS 8110)

To explain, in detail, the design of reinforced concrete structures requires a complete book in itself. However, the concepts behind simple reinforced concrete beams are easy to grasp with our current knowledge. We saw, in *chapter 3* that concrete is a brittle material with little tensile strength. It does, however, perform quite well in compression, although it is still less than 10% as strong as steel. In a reinforced concrete beam, the concrete is assumed to have zero tensile strength, the whole of the tensile force being resisted by the steel reinforcement. In fact the concrete on the tensile side of the beam is assumed to have cracked. *Figure 7.39* shows a steel 'I' beam and a rectangular reinforced concrete beam of roughly equivalent strength. A key difference between them is that the steel beam is reversible, whereas the concrete beam must operate with the reinforcement in the tensile face. A significant number of structural failures are caused by reinforcement being in the wrong face. In steel beams the web supports the shear force, whereas in concrete beams this is partly the job of the concrete and partly that of the **link** or **stirrup** bars, also shown in *figure 7.39*. The links are spaced at regular intervals throughout the length of the beam and, together with the small longitudinal bars in the compression zone, form a reinforcement cage.

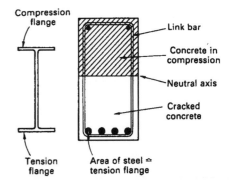

Figure 7.39
Steel and reinforced concrete beams of roughly equivalent strength

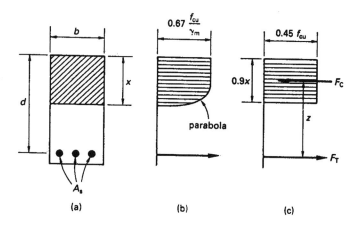

Figure 7.40
Stresses in a reinforced concrete beam at ultimate failure

7.9.1 *Reinforced concrete in bending*

Because reinforced concrete consists of two dissimilar materials it can fail in two ways. Either it can fail due to a lack of reinforcement – known as an **under-reinforced** failure. Or it can fail due to crushing of the concrete – an **over-reinforced** failure. Under-reinforced beams are preferred, as the plastic nature of the steel as it yields produces a beam which continues to support its load over large deformations. Over-reinforced beams, on the other hand, can fail rapidly as the relatively brittle concrete crushes.

At its ultimate moment, the distribution of stresses over the cross-section of a reinforced concrete beam is as shown in *figure 7.40b*. The shape of the concrete stress distribution is known as a **stress block**. For design purposes this can be simplified to the rectangular stress block shown in *figure 7.40c*. Eurocode 2 allows a similar simplification to *figure 7.40c* but the stress is $0.57 f_{cu}$ and the depth of the stress block is $0.8x$. The following definitions apply:

d = effective depth of reinforcement (measured from compressive face of concrete to centre of reinforcement)
b = width of concrete
x = neutral axis depth
z = lever arm (centre of compression to centre of tensile bars)
F_C = total compressive force
F_T = total tensile force
A_s = area of tensile reinforcement
f_{cu} = concrete characteristic strength

f_y = reinforcement characteristic strength
γ_m = partial safety factor for material strength (from *table 3.6*, γ_m for concrete = 1.5 and γ_m for steel reinforcement = 1.15)

We know that, to satisfy horizontal equilibrium, $F_C = F_T$ at all times. However, it is likely that either the concrete or the steel will reach ultimate collapse before the other. The ultimate moment that can be carried by the beam is therefore given by the **minimum** of the following two cases:

1. Based on the concrete in compression:

(ΣM about F_T) $M_{ult} = F_C \times z$

where $F_C = 0.67 f_{cu}/\gamma_m \times 0.9x \times b$
and $z = d - 0.45x$

Most designers limit the maximum depth of the neutral axis, x, to $0.5d$. If we substitute in the above for this and γ_m we get:

$$F_C = 0.444 f_{cu} \times 0.45d \times b$$
and $z = d - 0.225d = 0.775d$

Hence $M_{ult} = 0.444 f_{cu} \times 0.45d \times b \times 0.775d$
$$\mathbf{M_{ult} = 0.156 f_{cu} b d^2} \qquad [7.16]$$

2. Based on the steel in tension:

(ΣM about Fc) $M_{ult} = F_T \times z$

where $F_T = A_s \times f_y/\gamma_m = A_s f_y/1.15$
and conservatively $z = 0.775d$ as above

(This is conservative because, with an under-reinforced beam, the depth to the neutral axis, x, will be less than $0.5d$. This is because a smaller depth of concrete is sufficient to balance the tensile force from the reinforcement. Consequently z will be greater than $0.775d$.) This is dealt with in more detail in *section 7.8.3* which deals with slabs, however, the method is equally applicable to beams.

Hence $$M_{ult} = 0.674 A_s f_y d \qquad [7.17]$$

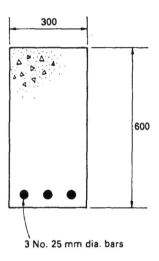

300

600

3 No. 25 mm dia. bars

Figure 7.41
A reinforced concrete beam

Example 7.12

Analyse the beam shown in *figure 7.41* to determine its ultimate moment capacity.

Concrete characteristic strength, f_{cu} = 30 N/mm^2
Steel characteristic strength, f_y = 460 N/mm^2
Concrete cover to main bars   = 40 mm

Solution

effective depth, d = Total depth – cover – 0.5 × bar dia.
$$d = 600 - 40 - 12.5 = 547.5 \text{ mm}$$

1. Based on concrete:
From *[7.16]*
$$\begin{aligned} M_{ult} &= 0.\,156f_{cu}bd^2 \\ &= 0.156 \times 30 \times 300 \times 547.5^2 \\ &= 420.9 \times 10^6 \text{ Nmm} \\ &= 420.9 \text{ kNm} \end{aligned}$$

2. Based on steel reinforcement:

$$\text{Area of steel, } A_s = 3 \times \pi \times 12.5^2 = 1473 \text{ mm}^2$$

From *[7.17]*
$$\begin{aligned} M_{ult} &= 0.674A_s f_y d \\ &= 0.674 \times 1473 \times 460 \times 547.5 \\ &= 250.0 \times 10^6 \text{ Nmm} \\ &= 250.0 \text{ kNm} \end{aligned}$$

M_{ult} must be limited to the lower of the above values.

Answer M_{ult} = 250.0 kNm

Comment – Because the limiting value of M_{ult} is based on the reinforcement and not the concrete area, this beam is underreinforced. This means that if a bending moment of 250.0 kNm was applied to the beam, the steel reinforcement would be on the point of yielding.

7.9.2 Reinforced concrete in shear

The actual behaviour of reinforced concrete under shear loading is very complex and still the subject of debate among researchers.

However, design techniques have been developed which work in practice. The first step is to convert the shear force, V, into an equivalent **design shear stress**, v, by dividing by the effective cross-sectional area of the beam:

$$\text{Design shear stress, } v = \frac{V}{b \times d}$$

The magnitude of the stress supported by the concrete alone is v_c, which depends upon the depth of the beam, the concrete strength and the percentage of tensile reinforcement. Values of v_c can be obtained from *table 7.1*. These values already contain a partial safety factor for material strength.

Table 7.1
Values of concrete shear strength,
v_c (N/mm^2)

$\dfrac{100 A_s}{bd}$	*Effective depth, d* (mm)	
	200	>=400
0.15	0.47	0.40
0.25	0.55	0.47
0.50	0.70	0.59
0.75	0.80	0.67
1.00	0.88	0.74
1.50	1.01	0.84
2.00	1.11	0.94
3.00	1.27	1.07

The above values are for grade 40 concrete.
For grade 30 concrete multiply by 0.909.
Based on BS 8 11 O: Part 1

To determine the shear force supported by the link reinforcement consider the free body diagram to one side of the crack, shown in *figure 7.42*. Clearly only the bars which actually cross the crack contribute. For a 45° crack:

$$\text{Number of bars} = d/s_v$$

$$\text{Shear force supported} = d/s_v \times f_{yv}/\gamma_m \times A_{sv}$$

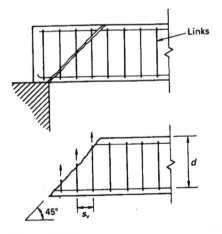

Figure 7.42
The free body to one side of a 45° shear crack

where:

s_v = spacing of links
A_{sv} = area of all legs of each shear link (here two)
f_{yv} = characteristic strength of shear links – usually mild steel bars with f_{yv} = 250 N/mm^2 are used
γ_m = 1.15 – partial safety factor for steel reinforcement strength.

Hence $f_{yv}/\gamma_m = 250/1.15 = 217$ N/mm^2

and

Shear force supported = $217A_{sv}d/s_v$

The above force is converted to an equivalent stress over the whole beam, v_s, by dividing by the cross-sectional area of the concrete, bd

$$v_s = \frac{217A_{sv}}{bs_v}$$

Summing the contributions from the concrete and steel we get:

$$v = v_c + v_s$$
$$= v_c + \frac{217A_{sv}}{bs_v}$$

Rearranging $\boldsymbol{A_{sv} = \frac{(v - v_c)bs_v}{217}}$ [7.18]

The above formula allows the size of mild steel shear links to be determined for a particular spacing.

In all but the most minor beams, a minimum area of links must be provided throughout the whole beam. This area is defined as that which will support a shear stress of 0.4 N/mm^2. So for mild steel links:

$$\text{Minimum } A_{sv} = \frac{0.4bs_v}{217} = \frac{bs_v}{543}$$ [7.19]

This means that whenever the shear stress on a beam is less than $(v_c + 0.4)$, at least minimum links must be used.

7.9.3 *Reinforcement spacing*

The production of detailed reinforcement drawings is beyond the scope of this book, however, there are some simple rules concerning the spacing of reinforcement that should be complied with. If bars are too closely spaced they impede the placing and compacting of the concrete, and hence voids can form. If bars are too widely spaced, excessively large cracks can occur.

Minimum spacing of bars = bar diameter or maximum size of aggregate plus 5 mm.

Maximum spacing of main bars = 300 mm for mild steel bars (f_y = 250 N/mm^2) and 160 mm for high yield bars (f_y = 460 N/mm^2).

Maximum spacing of shear links = 0.75 × beam effective depth.

Example 7.13

In *examples 4.8* and *7.5* we considered the design of steel beams to support the roof of a double garage. Beam B1 was subjected to a bending moment, M = 61.74 kNm, and a shear force, V = 41.2 kN. Design an alternative beam in reinforced concrete.

Given: f_{cu} = 40 N/mm^2, f_y = 460 N/mm^2, f_{yv} = 250 N/mm^2, cover = 20 mm.

Solution

Concrete dimensions

As the overall cross-sectional dimensions of the concrete beam are not prescribed, there is an infinite number of solutions to this problem. The aim is to produce a solution which is 'reasonable' and economic in practical terms.

A beam has reasonable proportions when its depth is twice its width. We can therefore use *equation [7.16]* to obtain an estimate of b and d.

$$M_{ult} = 0.156 f_{cu} b d^2$$

If $b = d/2$

$$M_{ult} = 0.0785f_{cu}d^3$$

But we require $M_{ult} = 61.74$ kNm

therefore $61.7 \times 10^6 = 0.0785 \times 40 \times d^3$
$$d = 270 \text{ mm}$$

Allowing for cover and the diameters of the link and main bars we can round up the sizes to 325 mm × 160 mm. If we assume 20 mm diameter main bars and 8 mm links:

$$d = 325 - 20 - 10 - 8 = 287 \text{ mm}$$

Main reinforcement

Equation [7.17] $M_{ult} = 0.674A_s f_y d$

Also $M_{ult} = 61.74$ kNm

Therefore $A_s = \dfrac{61.74 \times 10^6}{0.674 \times 460 \times 287}$

$$= 694 \text{ mm}^2$$

From *table 3.1* we can see that two 20 mm diameter bars plus one 10 mm diameter bar have the following area:

$$\text{Area} = (2 \times 314) + 78.5 = 706.5 \text{ mm}^2$$
$$> 694 \text{ mm}^2$$

Answer   Use two 20 mm dia. and one 10 mm dia. high yield bars

Shear reinforcement

$$\text{Design shear stress, } v = \frac{V}{b \times d} = \frac{41.2 \times 10^3}{160 \times 287}$$

$$= 0.90 \text{ N/mm}^2$$

$$\text{Reinforcement } \% = \frac{100A_s}{bd} = \frac{100 \times 706.5}{160 \times 287}$$

$$= 1.54$$

Interpolating from *table 7.1* we obtain $v_c = 0.91$ N/mm^2.

As the design shear stress, $v < (v_c + 0.4)$ minimum links apply. Place the links at 150 mm spacing:

from *equation [7.19]*

$$\text{Minimum A}_{sv} = \frac{bs_v}{543} = \frac{160 \times 150}{543}$$

$$= 44.2 \text{ mm}_2$$

From *table 3.1*, area of two 6 mm diameter bars = 56.6 mm^2

Answer   Use 6 mm dia. mild steel links at 150 mm spacing

The solution is shown in *figure 7.43*.

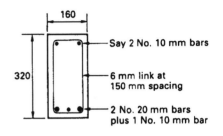

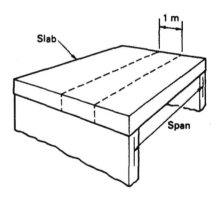

Figure 7.43
Solution to example

7.9.4 *Reinforced concrete slabs*

A solid reinforced concrete slab is designed as a shallow rectangular beam. The usual procedure is to consider a strip 1 m wide (*figure 7.44*). The depth of a slab is generally determined by the need to control its deflection, rather than by the strength of the concrete. Deflections are dealt with in more detail in *chapter 14*. A good starting point is to say that the effective depth will equal the span length divided by 20. To determine the area of reinforcement we could conservatively use *equation [7.17]* given in *section 7.8.1*:

$$M_{ult} = 0.674 \, A_s f_y d$$

The above equation is based on the assumption that the depth of concrete in compression, x, is equal to the maximum value of $d/2$. In fact with an under-reinforced beam or slab a smaller depth of concrete will provide sufficient 'push' to balance the 'pull' from the tensile reinforcement. This increases the lever arm, z, and hence reduces the amount of tensile reinforcement required, thus improving economy. From *figure 7.45* for a beam of width b and effective depth, d:

Taking moments about the tensile reinforcement

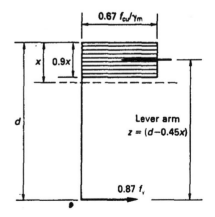

Figure 7.44
A one metre wide strip is considered in the design of slabs

Figure 7.45
The stresses in an under-reinforced beam or slab

$$M_{ult} = 0.67 f_{cu}/\gamma_m \times 0.9bx \times (d - 0.45x)$$
$$M_{ult} = 0.4 f_{cu}bdx - 0.18 f_{cu}bx^2 \qquad [7.20]$$

We can now solve the above quadratic equation for x. The lever arm is then obtained from $z = d - 0.45x$. An upper limit to the value of the lever arm, z, is $0.95d$. The reinforcement area is then evaluated from:

$$M_{ult} = zA_s f_y/\gamma_m$$
$$M_{ult} = 0.87\, zA_s f_y \qquad [7.21]$$

This method clearly has the disadvantage of the designer having to solve a quadratic equation for each beam or slab. To avoid this, design charts (*figure 7.46*) have been produced for different concrete and reinforcement strengths. The designer simply evaluates M_{ult}/bd^2 and then reads the value of $100\, A_s/bd$ off the chart. A wide range of charts has been produced to cover the design of almost all rectangular beams and slabs.

Once an area of tensile steel has been obtained, rather than converting it to a number of bars, it is more convenient for slabs to convert it to a particular bar size at a particular spacing. *Table 7.2* can be used for this purpose.

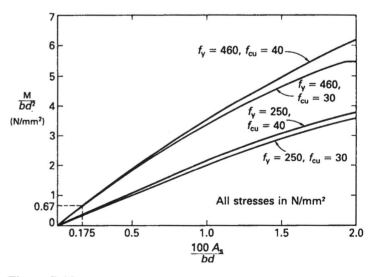

Figure 7.46
Beam design chart

Table 7.2
Reinforcement areas per metre width for various bar spacings (mm^2)

Bar size (mm)	Bar spacing (mm)								
	75	100	125	150	175	200	225	250	300
8	671	503	402	335	287	252	223	201	168
10	1047	785	628	523	449	393	349	314	262
12	1508	1131	905	754	646	566	503	452	377
16	2681	2011	1608	1340	1149	1005	894	804	670
20	4189	3142	2513	2094	1795	1571	1396	1257	1047

As well as the main reinforcing bars, which obviously run parallel to the direction of span, we must provide **secondary reinforcement**, which is perpendicular to the main bars. The two sets of bars are wired together to form a grid. The purpose of the secondary bars is to control cracking and to distribute concentrated loads. The minimum area of secondary reinforcement is given as a percentage of the total concrete area, as follows:

$$\text{for } f_y = 460 \text{ N/mm}^2 - \text{minimum area} = 0.13\% \text{ of } bh$$
$$\text{for } f_y = 460 \text{ N/mm}^2 - \text{minimum area} = 0.24\% \text{ of } bh$$

where b is the width being considered, and h is the overall thickness of the slab. So for a width of slab of 1 m the above areas become $1.3h$ mm^2 and $2.4h$ mm^2 respectively.

Shear is rarely a problem with solid simply supported slabs. The shear stress in the concrete will normally be less than the value of v_c from *table 7.1*, and in this case there is no need to provide shear reinforcement. If the shear stress does exceed v_c, shear reinforcement can be provided, but more generally the thickness of the slab is increased until the value of stress falls below that level.

To summarise, the steps involved in the design of a solid concrete slab are as follows:

Step 1 – Evaluate the effective depth of the slab by assuming that it is equal to span/20. Add on the required concrete cover, plus half a reinforcing bar diameter, to obtain the total depth.

Step 2 – Determine the total design load on the slab for one square metre.

Step 3 – Evaluate the maximum shear force and bending moment for a one-metre wide strip of slab.

Step 4 – Determine the required area of tensile reinforcement from either *equation [7.17]*, *equations [7.20]* and *[7.21]* or the design chart.

Step 5 – Check that the shear stress is less than v_c from *table 7.1*.

Step 6 – Convert the steel area to a particular bar size and spacing from *table 7.2*.

Step 7 – Design the secondary reinforcement.

Example 7.14

A simply supported concrete slab spans 4.5 m, and supports a characteristic imposed load of 3.0 kN/m². The characteristic dead load from finishes is 1.0 kN/m² in addition to the self-weight of the slab. Determine a suitable slab depth and arrangement of main and secondary reinforcement.

Unit weight of concrete = 24 kN/m³,
f_{cu} = 40 N/mm², f_y = 460 N/mm², cover = 20 mm

Solution

Step 1

$$\text{Effective depth, } d = 4500/20 = 225 \text{ mm}$$

Estimating the diameter of main bars at 16 mm

$$\text{Total depth, } h = 225 + 20 + 8 = 253 \text{ mm}$$

Try 250 mm thick slab (effective depth, d = 222 mm)

Step 2

$$\text{Dead load of finishes} = 1.0 \text{ kN/m}^2$$
$$\text{Dead load of stab} = 24 \times 0.25 = \underline{6.0 \text{ kN/m}^2}$$
$$\text{Total characteristic dead load} = 7.0 \text{ kN/m}^2$$
$$\text{Design load} = (1.4 \times 7.0) + (1.6 \times 3.0)$$
$$= 14.6 \text{ kN/m}^2$$

which for a one-metre wide slab corresponds to a UDL load of 14.6 kN/m.

Step 3

$$\text{Design bending moment} = wL^2/8 = 14.6 \times 4.5^2/8$$
$$= 37.0 \text{ kNm}$$
$$\text{Design shear force} = wL/2 = 14.6 \times 4.5/2$$
$$= 32.9 \text{ kN}$$

See *figure 7.47a* for the shear force and bending moment diagrams.

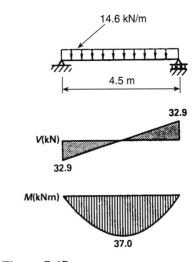

Figure 7.47a
Shear force and bending moment diagrams

Step 4

We will evaluate the required area of reinforcement by three methods and compare the results.

Method 1 – Using *equation [7.17]*

From
$$M_{\text{ult}} = 0.674 \, A_s f_y d$$

$$A_s = \frac{32.9 \times 10^6}{460 \times 0.674 \times 222} = 478 \text{ mm}^2$$

Method 2 – Using *equations [7.20]* and *[7.21]*

From *equation [7.20]*

$$M_{\text{ult}} = 0.4 f_{\text{cu}} bdx - 0.18 f_{\text{cu}} bx^2$$
$$32.9 \times 10^3 = 0.4 \times 40 \times 1000 \times 222x - 0.18 \times 40 \times x^2$$

Rearranging

$$0.72x^2 - 355.2x + 3290 = 0$$

solving $x = \dfrac{355.2 \pm \sqrt{(355.2^2 - 4 \times 0.72 \times 3290)}}{2 \times 0.72}$

$$x = 696.8 \text{ mm or } 9.44 \text{ mm}$$

The first of these is clearly not possible, as it is greater than the depth of the slab.

Therefore   lever arm, $z = d - 0.45x$
$$= 222 - (0.45 \times 9.44) - 217.8 \text{ mm}$$
but upper limit
$$= 0.95d = 0.95 \times 222$$
$$= 210.9 \text{ mm}$$

From [7.21] $M_{\text{ult}} = 0.87\, zA_s f_y$

$$A_s = \frac{32.9 \times 10^6}{0.87 \times 460 \times 210.9} = 390 \text{ mm}^2$$

Method 3 – Using the design chart

$$M/bd^2 = 32.9 \times 10^6 / 1000 \times 222^2$$
$$= 0.67$$

From *figure 7.46*

$$100\, A_s/bd = 0.175$$
$$A_s = 0.175 \times 1000 \times 222/100$$
$$= 389 \text{ mm}^2$$

Comment – It can be seen that the second two methods are in reasonable agreement. In this case method 1 produces more than 25% extra reinforcement compared to the other two methods.

Step 5

$$\text{Design shear stress, } v = \frac{V}{b \times d}$$

$$= \frac{32.9 \times 10^3}{1000 \times 222} = 0.148 \text{ N/mm}^2$$

From *table 7.1*

Shear strength, $v_c = 0.48$ N/mm$^2 > 0.148$ N/mm^2

Shear satisfactory

Step 6

Using $\qquad A_s = 389$ mm^2 and *table 7.2*

Answer   Use 12 mm dia. high yield bars at 275 mm spacing

$(A_s = 411$ mm$^2)$

Step 7

Area of secondary bars $= 1.3h$ mm^2
$$= 1.3 \times 250 = 325 \text{ mm}^2$$

Answer   Use 10 mm dia. high yield bars at 225 mm spacing

$(A_s = 349$ mm$^2)$

The solution is shown in *figure 7.47b*.

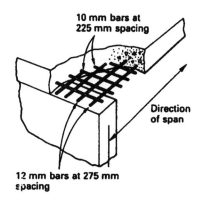

Figure 7.47b
A detail of the bars in the slab

7.10 Composite steel and concrete beams

Figure 7.2e showed a concrete slab supported by a steel beam. A considerable increase in both strength and stiffness occurs if an adequate connection is provided between the concrete and the steel. This means that for a sagging beam, the concrete will resist all or most of the compression and all of the tension will be resisted by the steel. This is known as **composite action** and such beams are often referred to as simply **composite construction**. (In fact, of course, both reinforced concrete and glass-reinforced polymers are also forms of composite construction, i.e. two different materials working in combination.) Composite steel and concrete structures are covered in Eurocode 4: BS EN 1994 and BS 5950: Part 3.

Composite construction provides significant economic savings

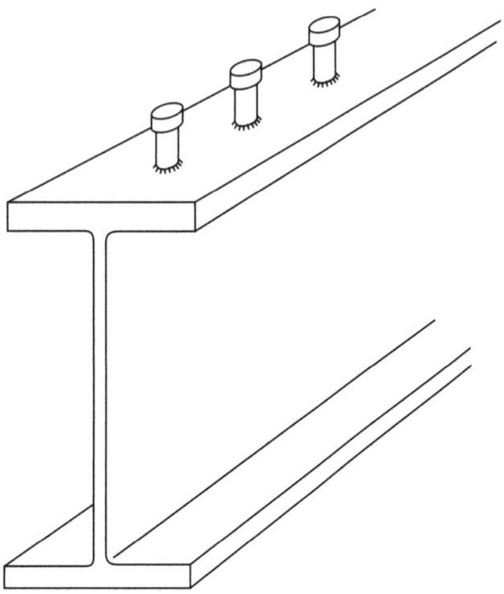

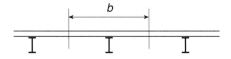

Figure 7.48
Shear studs welded to top of steel beam

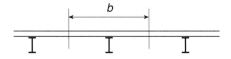

Figure 7.49
Effective width of concrete slab equals spacing of steel beams

compared to non-composite action where the concrete slab is simply regarded as a dead load imposed on the steel beam. Utilising the strength of the concrete in compression to supplement the steel beam can reduce the weight of steel required by 25%. The reinforced concrete slab is effectively providing a double benefit:

- It forms the slab that spans between the main beams, (i.e. at right angles to the main beams).
- It supplements the strength of the main beams.

In addition the steel beams can be used to support the formwork during the casting of the slab.

The connection between the steel beam and the concrete slab often takes the form of a row of steel studs welded to the top of the steel flange, as shown in *figure 7.48*.

7.10.1 *Design procedure for composite beams*

The first step is to determine the dimensions of the composite section. The thickness of the slab is determined by the requirement to span between the main beams and the procedure shown in *section 7.9.4* for the design of slabs applies. It should be pointed out, however, that the slab will not generally be simply supported between the steel beams. It will usually consist of several continuous spans as shown in *figure 7.49*. It is thus a statically indeterminate structure, however, in practice the in-span and support bending moments are usually determined from simple standard cases. For approximately equal spans (i.e. within 15%) the values shown in *table 7.3* are typical.

Table 7.3
Bending moments in continuous slabs

	End span	Penultimate support	Interior span	Interior support
Dead load	$+\dfrac{wl^2}{12}$	$-\dfrac{wl^2}{10}$	$+\dfrac{wl^2}{24}$	$-\dfrac{wl^2}{12}$
Imposed load	$+\dfrac{wl^2}{10}$	$-\dfrac{wl^2}{9}$	$+\dfrac{wl^2}{12}$	$-\dfrac{wl^2}{9}$

The above clearly indicate a reduction compared to the simply supported value of $wl^2/8$. The support moments are hogging and so, of course, require reinforcement in the top of the slab, whereas the span moments are sagging and require it in the bottom face.

The effective width of the concrete slab is generally taken to be the centres of the steel beams. The steel beam size must initially be estimated, resulting in the complete trial section as shown in *figure 7.50*. This enables the ultimate strength of the composite section to be determined and compared with the required value. Several iterations may be required to obtain the optimum beam size.

As with reinforced concrete beams in *section 7.9.1*, the ultimate compressive strength of concrete is assumed to be $0.45\,f_{cu}$. The ultimate tensile and compressive strength of the steel beam is the yield stress, σ_y. Also, as with the reinforced concrete beam, the total compressive force on the beam cross-section, F_C, must equal the total tensile force, F_T, if horizontal equilibrium is to be satisfied. Two cases are possible:

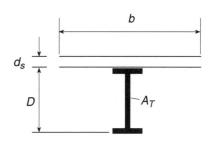

Figure 7.50
Cross-section of composite beam

1. The neutral axis lies within the depth of the concrete slab. This occurs when the ultimate compressive strength of the concrete exceeds the ultimate tensile strength of the steel beam. In this case all concrete below the neutral axis is ignored.

2. The neutral axis is in the steel beam. This occurs when the ultimate tensile strength of the steel beam is greater that the ultimate compressive strength of the concrete slab. In this case the portion of the steel beam above the neutral axis supplements the concrete slab in providing compression.

In the above:

Ult. compressive strength of concrete, $F_{CC} = 0.45f_{cu}bd_s$
Ult. tensile strength of steel $\qquad\qquad = A\sigma_y$

Case 1 – Neutral axis in slab – see figure 7.51a
The depth to the neutral axis, x, is determined so that the compressive strength of the concrete above the axis is equal to the total tensile strength of the steel beam:

$$bx \times 0.45f_{cu} = A \times \sigma_y$$

$$\text{neutral axis depth, } x = \frac{A\,\sigma_y}{0.45f_{cu}b} \qquad [7.22]$$

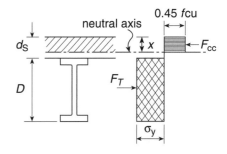

Figure 7.51a
Neutral axis within concrete slab

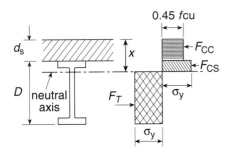

Figure 7.51b
Neutral axis within steel beam

We can now determine the ultimate resistance moment by taking moments about the centre of either tension or compression:

$$(\Sigma M \text{ about } F_C) \qquad M_{\text{ult}} = A \times \sigma_y \times \left(\frac{D}{2} + d_s - \frac{x}{2}\right) \qquad [7.23]$$

Case 2 – Neutral axis in steel beam – see figure 7.51b
In this case the compressive force is made up of two components, one from the concrete slab and one from the portion of steel beam above the neutral axis:

$$F_C = F_{CC} + F_{CS}$$

Where F_{CC} = compressive force in concrete slab
         F_{CS} = compressive force in steel beam

The first step is to evaluate the concrete slab component of compression, F_{CC}:

$$F_{CC} = 0.45 f_{\text{cu}} \, bd_s$$

Next we determine the compression component from the steel beam above the neutral axis, F_{CS}. Because the yield stress of the steel beam is the same in both tension and compression we know that the total sum of tensile and compressive forces is given by:

$$F_C + F_T = F_{CC} + A\sigma_y$$

but $F_C = F_T$ for horizontal equilibrium

hence $2F_C = F_{CC} + A\sigma_y$

$$F_C = \frac{F_{CC} + A\gamma_y}{2}$$

From $F_{CS} = F_C - F_{CC}$

$$F_{CS} = \frac{F_{CC} + A\sigma_y}{2} - F_{CC} \qquad [7.24]$$

Having obtained the magnitude of the compressive force carried by the steel beam, F_{CS}, we now need see if the neutral axis lies within the top flange of the beam or occurs lower down in the beam web. The ultimate force capacity of the steel beam flange is given by:

$$\text{ultimate flange force} = \text{flange area} \times \sigma_y$$

If the above ultimate flange force is greater than F_{CS}, then the neutral axis occurs within the flange, at a depth below the top of the steel beam of:

$$\text{depth to neutral axis} = \frac{F_{CS}}{\text{flange width} \times \sigma_y} \qquad [7.25]$$

If the ultimate flange force is less than F_{CS}, then the neutral axis occurs in the web, at a depth below the top of the steel beam of:

$$\text{depth to neutral axis} = \text{flange thickness} + \frac{F_{CS} - \text{ult. flange force}}{\text{web thickness} \times \sigma_y}$$
$$[7.26]$$

The problem now is to find the ultimate resistance moment. With the previous case, where the neutral axis is in the concrete, this is easy because we know where both the centre of compression and the centre of tension lie. However, when the neutral axis lies in the steel this is not obvious. We could either calculate the location of the centroid of F_C and F_T (in a similar way to that shown in *section 7.7.4*) or consider the resistance moment to be made up of individual rectangular components as shown in *figure 7.53*. We can then take moments about the centre of the concrete slab to evaluate the ultimate resistance moment. This is best done in tabular form as shown in *example 7.15*.

Spacing of shear connectors
Shear connectors are required to join the concrete slab to the top flange of the beam. Relative movement between the bottom of the concrete and the top of the steel must be prevented if full composite action is to be generated. The commonest form of shear connection is a row of **headed studs** as shown in *figure 7.48*. The shear capacities of single studs, 100 mm long, are given in *table 7.4*.

The force to be transmitted across the interface is directly related to the rate of change of force in the concrete slab. This force is greatest at the point of maximum bending moment, and, for simply supported beams, reduces to zero at the supports. The common procedure for such beams is:

- determine the maximum force in the slab at the point of maximum bending moment

Table 7.4
Shear capacity of headed studs (kN)

Stud diameter (mm)	Concrete characteristic strength, f_{cu} (N/mm^2)		
	25	30	40
25	117	123	134
22	95	101	111
19	76	80	87

- select a stud size and determine the number of studs required to resist the above maximum force
- distribute the studs equally over the distance between the point of maximum bending moment and the support.

The minimum spacing between adjacent studs is five times the stud diameter although a double row can be used.

Note:
In theory, the force to be transmitted between the concrete slab and the steel beam is proportional to the rate of change of bending moment. We saw in section 7.3.5 that this rate of change equals the shear force at any point in the beam. For a simply supported beam with a uniformly distributed load we know that the shear force increases linearly towards the supports. We can therefore conclude that, consequently, the shear studs should get closer together towards the supports. The justification for an equal spacing, as described in the above procedure, is that at ultimate collapse the studs would yield near the supports and then redistribute the load to the studs nearer the centre of the span. Section 10.6 shows how an optimum variable spacing could be calculated.

Example 7.15

A proposed floor for a heavy factory consists of a 125 mm thick reinforced concrete slab spanning 2.0 m between a series of 533 × 210 × 122 kg/m universal beams. The main steel beams span 15 m and are simply supported. The weight of any finishes can be ignored.

a) Determine the design bending moment on one of the main beams.
b) Check that the ultimate resistance moment of the composite steel and concrete section is adequate.
c) Determine the spacing of suitable shear connectors.

Given:

$$f_{cu} = 30 \text{ N/mm}^2$$
$$\sigma_y = 275 \text{ N/mm}^2$$
Unit weight of concrete $= 24 \text{ kN/m}^3$
Characteristic imposed load $= 10 \text{ kN/m}^2$

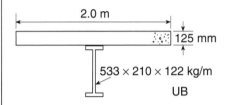

Figure 7.52
Cross-section of beam for *example 7.15*

Solution

a)
A cross-section of the composite beam is shown in *figure 7.52*.

From *Appendix* for $533 \times 210 \times 122$ kg/m universal beam:
$A = 156.0 \text{ cm}^2$, $D = 544.6$, $B = 211.9$ mm, $T = 21.3$ mm, $t = 12.8$ mm

Loads

Dead load of slab $= 24 \times 0.15$ $= 3.6 \text{ kN/m}^2$

Slab load/m of beam $= 3.6 \times 2.0$ $= 7.2 \text{ kN/m}$
Self-weight of steel beam $= 0.122 \times 9.81 = \underline{1.2 \text{ kN/m}}$
Total characteristic dead load $= 8.4 \text{ kN/m}$

Imposed load/m of beam $= 10.0 \times 2.0$ $= 20.0 \text{ kN/m}$

Design load $= (1.4 \times 8.4) + (1.6 \times 20.0)$
$= 43.8 \text{ kN/m}$

Design bending moment $= \dfrac{wL^2}{8} = \dfrac{27.76 \times 15^2}{8}$

Answer = 1230.8 kNm

b)

Ult. compressive strength of concrete

$$F_{CC} = 0.45 f_{cu} \, bd_s$$
$$= 0.45 \times 30 \times 2000 \times 125$$
$$= 3.38 \times 10^6 \text{ N}$$

Ult. tensile strength of steel

$$= A\sigma_y = 156.0 \times 10^2 \times 275$$
$$= 4.29 \times 10^6 \text{ N}$$

Because $A\sigma_y > 0.45 f_{cu} \, bd_s$ the neutral axis must occur below the slab in the steel beam. The compression component carried by the steel beam is:

From [7.24]
$$F_{CS} = \frac{F_{CC} + A\sigma_y}{2} - F_{CC}$$

$$= \frac{3.38 \times 10^6 + 4.29 \times 10^6}{2} - 3.28 \times 10^6$$

$$= 0.455 \times 10^6 \text{ N}$$

ultimate flange force = flange area × σ_y
$$= 211.9 \times 21.3 \times 275$$
$$= 1.24 \times 10^6$$

Because ult. flange force > F_{CS} the neutral axis must occur within the flange.

From [7.25]

$$\text{depth to neutral axis} = \frac{F_{CS}}{\text{flange width} \times \sigma_y}$$

$$= \frac{0.455 \times 10^6}{211.9 \times 275} = 7.81 \text{ mm}$$

We can now take moments about the centre of the concrete slab to find the ultimate resistance moment of all the rectangular components of the beam. *Figure 7.53* shows an exaggerated view of the composite cross-section showing the distances from the centre of the concrete slab to the centre of each rectangular element. The calculation is best done in tabular form as shown below. The steel is all at the yield stress, σ_y, however, that above

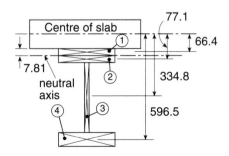

Figure 7.53

Distances from centre of concrete slab to centre of rectangles in mm the neutral axis (item 1) is in compression and that below it (items 2, 3, and 4) in tension. The area of item 1 is thus shown negative in the table.

Item	Area, A_n	lever arm, l_n	$A_n \times l_n \times 10^6$
1	-211.9×7.81	66.4	-0.11
2	211.9×13.49	77.1	0.22
3	12.8×502	334.8	2.15
4	211.9×21.3	596.5	$\underline{2.69}$
			4.95

Ultimate resistance moment $= 4.95 \times 10^6 \times 275$ Nmm

$$= 1361 \text{ kNm} > 1230.8 \text{ kNm}$$

Answer   Composite beam is adequate

c)
Using 22 mm diameter studs. From *table 7.4*, for grade 30 concrete, shear capacity $= 101$ kN

Max. force in slab $F_{CC} = 3.38 \times 10^6$ N

$$\text{Number of studs required} = \frac{3.38 \times 10^3}{101} = 34$$

$$\text{Spacing of studs} = \frac{7500}{34}$$

Answer   spacing = 220 mm

7.11 Summary of key points from chapter 7

1. The common types of beam are **simply supported** (pin at one end and roller at the other) and **cantilever** (built in at one end and free at the other).
2. If a beam is 'cut' and the equilibrium of the remaining free body to one side of the cut is considered, it is necessary to impose a **shear force** and a **bending moment** at the cut to restore equilibrium.
3. If values of shear force and bending moment are plotted

throughout the length of the beam the **shear force diagram** and the **bending moment diagram** result. These are vitally important in the design of beams.

4. One face of a beam has tensile stresses whereas the other is in compression. Our convention is always to plot bending moments on the **tensile** face of the beam.

5. For a simply supported beam of span L with a point load, P at mid-span, the maximum bending moment is **$PL/4$.**

6. For a simply supported beam of span L with a uniformly distributed load, w (UDL) throughout, the maximum bending moment is **$wl^2/8$**.

7. Throughout a uniformly distributed load, shear force diagrams slope and bending moment diagrams are curved.

8. The peak bending moment always occurs at a point of zero shear.

9. Bending stresses are zero on the neutral axis of a beam which passes through the centroid of the beam cross-section during elastic bending.

10. In limit state design it is generally the aim to ensure that a beam has an adequate safety factor against reaching its **fully plastic bending strength**, M_{ult}, where, for a steel beam:

$$M_{ult} = \text{yield stress}, \sigma_y \times \textbf{plastic section modulus}, S.$$

11. In elastic design the relationship between bending moment and stress is as follows:

$$M = \text{maximum stress}, \sigma_{max} \times \textbf{elastic section modulus}, Z.$$

12. Both the **plastic section modulus**, S, and the **elastic section modulus**, Z, are geometric properties of a particular beam cross-section, and can be calculated or looked-up in tables for standard beams.

13. If the top flange of a beam is not restrained against lateral buckling, the fully plastic bending strength must be reduced.

14. With an 'I' beam it is usual to assume that the shear force is resisted only by the web.

$$\text{Average shear force} = \frac{V}{Dt}$$

15. Concrete is weak in tension so beams must be reinforced with steel on the tensile face. The ultimate strength is the minimum

of either the moment to cause the concrete in the compression zone to crush:

$$M_{ult} = 0.156 f_{cu} bd^2$$

or the moment to cause tensile failure of the reinforcement:

$$M_{ult} = 0.674 A_s f_y d \text{ (conservative)}$$

For a more economic solution to the latter either solve a simultaneous equation or use a design chart.

16. Links (or stirrups) are provided to resist shear in reinforced concrete beams. For mild steel links the area required, A_{SV} for a particular spacing, s_V is given by:

$$A_{SV} = \frac{(v - v_c)bs_v}{217}$$

17. Reinforced concrete slabs are designed as beams of unit width and must also be provided with secondary bars perpendicular to the direction of span.

18. Composite construction refers to a concrete slab that is connected to a steel beam with shear connectors so that there is no relative movement under bending. The combination can reduce the required size of steel beam by 25%.

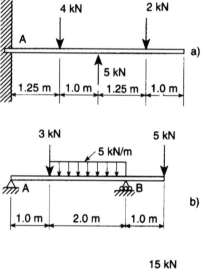

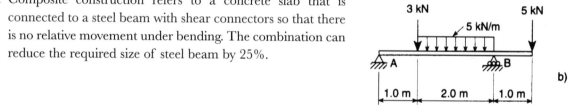

7.12 Exercises

E7.1   For each of the beams shown in *figure 7.54a–d* draw the shear force and bending moment diagrams. Indicate the magnitude of all peak values.

a) $M_{sag} = 0.5\ kNm$, $M_{hog} = 2.5\ kNm$
b) $M_{sag} = 3.71\ kNm$, $M_{hog} = 5\ kNm$
c) $M_{hog} = 30\ kNm$
d) $M_{sag} = 11.3\ kNm$, $M_{hog} = 4.5\ kNm$

E7.2   A simply supported beam spans 6.0 m and carries a design uniformly distributed load (UDL) of 12.0 kN/m. The top flange is fully restrained against lateral torsional buckling. Select a suitable mild steel standard universal beam.
use 406 × 140 × 39 UB

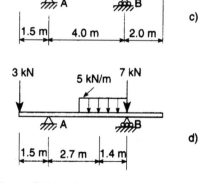

Figure 7.54a–d

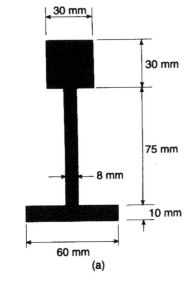

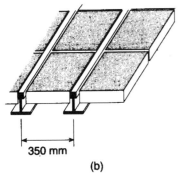

Figure 7.55

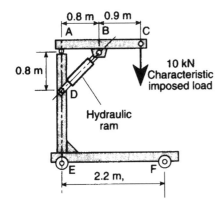

Figure 7.56

E7.3   *Figure 7.55a* shows the cross-sectional dimensions of steel beams which are placed at 350 mm centres to support removable floor panels in a computer room – see *figure 7.55b*. The whole of the floor area is subject to the following characteristic loads:

Characteristic dead load = 0.9 kN/m^2
Characteristic imposed load = 3.5 kN/m^2

a) Calculate the plastic section modulus for the beam.
b) Determine the design uniformly distributed load for a typical beam (in kN/m).
c) Calculate the maximum design span of a beam (σ_y = 275 N/mm^2).
a) S = 81.3 cm^3
b) w = 2.4 kN/m
c) l_{max} = 8.63 m

E7.4   A garage crane used to lift car engines is shown in *figure 7.56*. The jib ABC is lifted by means of the hydraulic ram BD. The characteristic imposed load to be lifted is 10kN. All other loads can be neglected.

a) Convert the characteristic imposed load into a design load.
b) Calculate the design load in the hydraulic ram and the axle loads at E and F when the jib is horizontal as shown.
c) Draw the bending moment diagram for the jib, and determine the dimensions of a suitable square hollow section (SHS) assuming that the jib is made from mild steel (σ_y = 275 N/mm^2). Ignore the effect of axial forces in the jib.
a) Design load = 16 kN
b) F_{ram} = 48.1 kN, R_E = 3.64 kN, R_F = 12.36 kN
c) Use 100 × 100 × 4.0 SHS

E7.5   *Figure 7.57* shows the cross-section of a mechanical part which is in bending. Determine the elastic stresses at the top and bottom of the section when subjected to a sagging bending moment of 0.5 kNm
σ_{top} = 153 N/mm^2, σ_{bottom} = 131 N/mm^2

E7.6   A rectangular reinforced concrete beam has overall cross-sectional dimensions of 450 mm deep and 200 mm wide and contains two 25 mm diameter tensile reinforcing bars which have 30 mm cover.

$$\text{Concrete characteristic strength, } f_{cu} = 30 \text{ N/mm}^2$$
$$\text{Steel characteristic strength, } f_y = 460 \text{ N/mm}^2$$

It is subject to a total uniformly distributed design load of 40.0 kN/m.

a) What is the maximum design span of the beam?
b) Determine a suitable arrangement of shear reinforcement at this maximum span.

a) Maximum span = 4.98 m
b) Shear reinforcement say 10 mm diameter mild steel links at 300 mm spacing

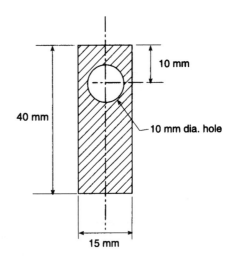

Figure 7.57

Chapter 8

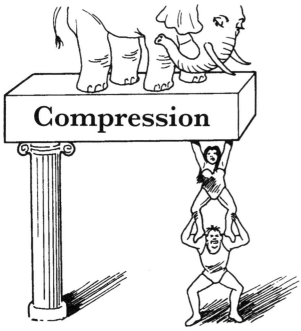

Compression

Topics covered

Examples of compression members
Short axially loaded columns
Slender columns
The design of compression members

8.1 Introduction

This chapter is concerned with the design of structural members subjected to compressive forces. A small compression member, such as that in a framed structure, is known as a **strut**. A larger member, such as the main support for a beam in a building, is known as a **column**, or more traditionally a **stanchion**.

Axially loaded compression members can fail in two principal ways:

- Short fat members fail by crushing or splitting of the material. This is a strength criterion.
- Long thin members fail by sideways buckling. This is a stiffness criterion.

Most structural members either lie in the long thin category, or somewhere in-between the two extremes where the behaviour can get very complex. Compression members subjected to eccentric loads or bending moments can also fail by local crushing of the material, but that is beyond the scope of this book.

8.2 Examples of compression members

Figure 8.1a shows the end of a steel roof truss. The sloping compression member (or **rafter**) is formed from two parallel angle sections fixed each side of the **gusset plates**. These are known as **back-to-back** angles, and it can be seen that the two angles are also connected together at regular intervals throughout their length. This increases the load-carrying capacity of the angles as they are less likely to buckle if joined together.

Figure 8.1b shows a typical **universal column** section at the corner of a steel-framed building. Whereas beams are usually 'I' shaped for maximum efficiency, columns are 'H' shaped – the aim being to get an approximately equal second moment of area, and hence stiffness, in all directions. Otherwise the column would invariable fail by buckling about its weaker axis.

A particularly efficient shape for compression members is the hollow box section (*figure 8.1c*). This is available in steel, aluminium alloy and fibre composite. However, it is more difficult

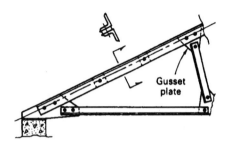

Figure 8.1a
The end of a steel roof truss with a top member (rafter) in compression

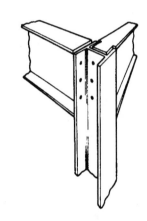

Figure 8.1b
The corner column of a steel-framed building

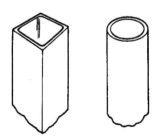

Figure 8.1c
Hollow sections make efficient compression members

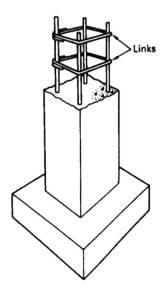

Figure 8.1d
A reinforced concrete column

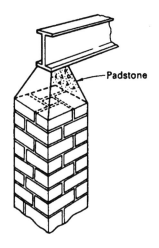

Figure 8.1e
A brickwork column

to connect to than open sections such as angles and universal columns. The circular hollow section is the most efficient shape of all, but the ratio of the overall tube diameter to the wall thickness should be kept below about 40 to prevent local crushing.

A reinforced concrete column, with part of the concrete removed, is shown in *figure 8.1d*. The compressive strength of the concrete is supplemented by the four corner steel reinforcing bars. Steel link bars must be fixed at regular intervals throughout the column to restrain the corner bars, and prevent them from bursting out of the concrete when fully loaded. The corner bars must also be properly anchored into the foundation at the bottom, so that the load is suitably dispersed.

Figure 8.1e shows a brickwork column. As with all brickwork structures, the bricks must be properly bonded so that there are no continuous vertical joints between two adjacent courses. Where concentrated loads from steel beams bear onto brickwork, it is often necessary to provide concrete **padstones** to spread the load.

8.3 Short axially loaded columns

A rectangular column is described as 'short' when its height to minimum width ratio is less than 15 if the top is restrained against lateral movement and less than 10 if unrestrained. In practice it is usually only reinforced concrete or brickwork columns that fall into this category. The design procedure is similar to that adopted for tension members, i.e. ensure that the cross-sectional area of the member is adequate to reduce the ultimate stress to an acceptably low value. *Example 1.1, part b* shows the process of designing a short column in brickwork.

8.3.1 *Short axially loaded columns in reinforced concrete*

If we consider the rectangular column cross-section shown in *figure 8.2*, we would expect the ultimate compressive load capacity, N, to be the sum of the strengths of both the concrete and steel components:

$$N = f_{cu} A_c + f_y A_{sc}$$

where

f_{cu} = characteristic concrete cube crushing strength
A_c = area of concrete
f_y = characteristic yield stress of steel
A_{sc} = area of steel.

However, as with beams, the cube test overestimates the strength of the concrete because of the restraint provided by the steel plattens of the testing machine. The equation is therefore empirically modified to become:

$$N = 0.67f_{cu} A_c + f_y A_{sc}$$

We must now divide by the partial safety factors for material strength, where γ_m, where γ_m = 1.5 for concrete and 1.15 for steel reinforcement. The equation now becomes:

$$N = 0.45f_{cu} A_c + 0.87f_y A_{sc}$$

Finally an allowance must be made for the fact that the load may not be applied purely axially. The equation is consequently reduced to:

$$N = 0.4f_{cu} A_c + 0.75f_y A_{sc} \qquad [8.1]$$

The steel reinforcement area must be between 0.4% and 6% of the concrete area. This clearly gives considerable scope for variation in the overall dimensions of the column. The designer could go for a small column with a large steel area or vice versa. In practical terms all the columns in a particular building are likely to have the same external dimensions, but the area of steel can be varied to compensate for varying loads.

The links shown in *figure 8.1d* should have a minimum diameter of one-quarter of that of the largest longitudinal bars, and the maximum spacing of the links should be 12 times the diameter of the smallest longitudinal bars.

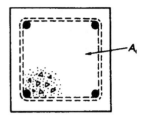

Figure 8.2

The cross-section of a reinforced concrete column

Example 8.1

A short reinforced concrete column is to support the following axial loads:

characteristic dead load = 758 kN
characteristic imposed load = 630 kN

If the column is to measure 325 mm × 325 mm and the concrete characteristic strength is 30 N/mm², determine the required size of high yield reinforcing bars, and specify suitable links.

Solution

First we must find the ultimate design load by applying the partial safety factors for load.

$$\text{Design load} = 1.4G_k + 1.6Q_k$$
$$= 1.4 \times 758 + 1.6 \times 630$$
$$= 2069 \text{ kN}$$

From *[8.1]* $N = 0.4f_{cu}A_c + 0.75f_yA_{sc}$

hence $2069 \times 10^3 = 0.4 \times 30 \times 325^2 + 0.75f_yA_{sc}$

$$801\,500 = 0.75 \times 460 \times A_{sc}$$
$$A_{sc} = 2323 \text{ mm}^2$$

using 4 bars area/bar = 581 mm²

From *table 3.1*:

$$\text{Area of 32 mm dia. bar} = 804 \text{ mm}^2$$

Comment – this is clearly somewhat over-size, but the next bar down, which is 25 mm diameter, only has an area of 491 mm² which is not adequate. An alternative solution would be to use eight 20 mm diameter bars which have a total area of 2513 mm².

$$\text{Steel percentage} = \frac{4 \times 804 \times 100}{325^2} = 3.04\%$$

This is between 0.4% and 6% and therefore satisfactory

$$\text{Minimum diameter of links} = 32/4 = 8 \text{ mm}$$
$$\text{Maximum spacing of links} = 32 \times 12 = 384 \text{ mm}$$

Answer   **Use four 32 mm dia. bars with 8 mm dia. links at 350 mm spacing**

8.4 Slender columns

We have seen above that the strength of a short column is based only upon its cross-sectional area and the material strength. For

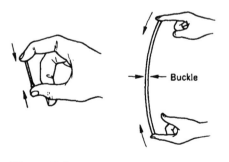

Figure 8.3

Increasing the length of a strut reduces its buckling load

slender columns this is not the case. A slender column fails by side-ways buckling, and common-sense tells us that the **length** of the column makes a significant difference. Thus, as shown in *figure 8.3*, a matchstick is reasonably strong in compression, but a 300 mm long stick of the same cross-sectional area would be very weak. The load at which a slender column buckles is known as its **critical buckling load, P_{crit}**.

8.4.1 *The Euler critical buckling formula*

Consider an object such as a plastic ruler under a small compressive load significantly less than P_{crit} (*figure 8.4a*). If we now apply a small sideways load as shown in *figure 8.4b* it will deflect sideways in the middle, but when the sideways load is removed it will spring straight again (*figure 8.4c*). If the compressive load is steadily increased, the point will come when the ruler does not spring back (*figure 8.4d*). This compressive load is P_{crit}. The moment induced by the eccentricity of P_{crit} just balances the restoring moment from the curvature of the beam. If the load is increased beyond P_{crit} collapse will occur. The value of the critical buckling load for a slender pin-ended column is given by the **Euler buckling formula**:

$$P_{crit} = \frac{\pi^2 EI}{L^2} \qquad [8.2]$$

where

$$E = \text{modulus of elasticity}$$
$$I = \text{second moment of area}$$
$$L = \text{length between pins.}$$

This formula is very important as it forms the basis of the design methods for real compression members given later in this chapter. Note that it does not contain a stress value. The *EI* term is a measure of bending stiffness. It also shows clearly how the critical buckling load is inversely proportional to the square of the length. The formula is not easy to derive, but the derivation is given below for those who are interested.

Figure 8.4a
A ruler with a small compressive load, P

Figure 8.4b
A horizontal load is applied

Figure 8.4c
The horizontal load is removed and the ruler returns to straight

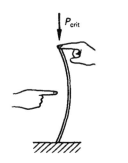

Figure 8.4d
If P is increased to P_{crit} the ruler remains bent

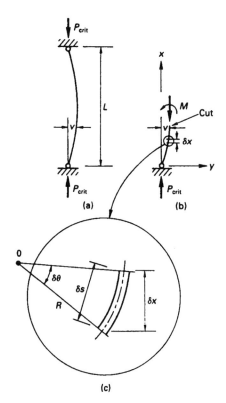

Figure 8.5
Derivation of the formula for the buckling of a strut

Proof of the Euler buckling formula

We shall use the following formula from the engineer's equation of bending – *equation [7.9]*:

$$\frac{M}{I} = \frac{E}{R}$$

Figure 8.5a shows a pin-ended strut of length L supporting its critical buckling load, P_{crit}. Assume the horizontal deflection at any point is y. First cut the strut at any arbitrary point and consider the equilibrium of the free body formed from the lower part (*figure 8.5b*). Now consider a small element of the strut of length δs (*figure 8.5c*), where the radius of curvature is R:

For θ in radians
$$R\delta\theta = \delta s$$

Therefore
$$\frac{1}{R} = \frac{d\theta}{ds} \qquad [8.3]$$

For a small lateral deflection, $\delta s = \delta x$ and as the length of the element approaches zero we can replace δx with the differential.

[8.3] becomes
$$\frac{1}{R} = \frac{d\theta}{dx}$$

Also
$$\text{slope, } \theta = \frac{dy}{dx}$$

Therefore
$$\frac{1}{R} = \frac{d^2y}{dx^2} \qquad [8.4]$$

And from *[7.16]* above:

$$M = \frac{EI}{R} = EI\frac{d^2y}{dx^2} \qquad [8.5]$$

From *figure 8.5b* take moments about the cut:

$$P_{crit} \times (-y) = EI\frac{d^2y}{dx^2} \qquad [8.6]$$

(y is negative relative to the cut)

$$\frac{d^2y}{dx^2} + \frac{P_{crit}y}{EI} = 0 \qquad [8.7]$$

The above is a second-order, constant-coefficient, linear homogeneous differential equation! The solution is as follows:

Substituting k^2 for P_{crit}/EI we get:

$$\frac{d^2y}{dx^2} + k^2y = 0$$

Auxiliary equation:

$$m^2 + k^2 = 0$$

therefore
$$m = \pm \, ik$$
$$y = Ae^{ikx} + Be^{-ikx}$$

or
$$y = A(\cos kx + i\sin kx) + B(\cos kx + i\sin kx)$$
$$= (A + B)\cos kx + i(A - B)\sin kx$$

We are only interested in a real solution and the coefficients become real if A and B are complex conjugates:

Let
$$A = a + ib$$
$$B = a - ib$$

so
$$A + B = 2a = C_1$$
$$i(A - B) = -2b = C_2$$

hence
$$\boldsymbol{y = C_1\cos kx + C_2\sin kx} \qquad [8.8]$$

Comment – Try differentiating [8.8] twice and then substituting into [8.7] to show that it is a valid solution.

The boundary conditions for evaluating the constants are:

at
$$x = 0 \qquad y = 0$$
and at
$$x = L \qquad y = 0$$

Substituting the above values into *[8.8]* we get:

$$C_1 = 0$$
and
$$C_2 \sin kL = 0$$

This equation is satisfied when $kL = 0, \pi, 2\pi, \ldots$, etc.

Obviously when either k or L are zero the solution is of no interest (implies zero load or zero length). However, when $kL = \pi$ we get:

$$k = \frac{\pi}{L} \text{ and } k^2 = \frac{\pi^2}{L^2}$$

Substituting for k $$\boldsymbol{P}_{\textbf{crit}} = \frac{\pi^2 \boldsymbol{EI}}{\boldsymbol{L^2}}$$ [8.9]

Solutions with $kL - 2\pi$, 3π etc. produce higher loads which are of no practical interest as the strut would already have buckled under the above load. Finally the equation is also satisfied when $C_2 = 0$. This implies that there is no deflection and that the strut remains perfectly straight. This is a case of **unstable equilibrium**, like balancing a pencil on its point.

Example 8.2

Determine the critical Euler buckling load of a solid aluminium rod of diameter 12 mm if it has a length of 1.0 m between pin-ended supports.

Solution

From *[8.9]* $$P_{\text{crit}} = \frac{\pi^2 EI}{L^2}$$

From *table 3.4* $E = 70\,000 \text{ N/mm}^2$
From *figure 7.31* $I = \pi r^4/4 = \pi \times 6^4/4 = 1018 \text{ mm}^4$
    $L = 1000 \text{ mm}$

Therefore $$P_{\text{crit}} = \frac{\pi^2 \times 70\,000 \times 1018}{1000^2} = 703 \text{ N}$$

Answer   Critical Euler buckling load = 0.703 kN

8.4.2 *The design of real compression members*

The Euler buckling formula works reasonably well for very slender struts, however, it over-estimates the strength of shorter compression members. It is in fact possible for the formula to indicate a critical buckling load which would cause the stress in the member to be above the yield stress for the material, and this is clearly not acceptable. Also real struts cannot be assumed to be initially perfectly straight, or loaded perfectly axially. In national standards these factors are taken into account by using formulae which are modified as a result of considerable empirical testing in the laboratory. The **Perry–Robertson formula** is commonly used, and the results are usually presented in the form of curves or tabulated values of compressive strength, p_c, against **slenderness ratio**.

$$\text{Slenderness ratio} = \frac{\text{effective length}}{\text{radius of gyration}} = \frac{L_E}{r}$$

For a pin-ended member the effective length, L_E, is the distance between the pins. Members with other types of end restraint are dealt with in the next section. The radius of gyration was defined at the end of *chapter 7*, and is given in section tables for standard sections. Where more than one value for r is given in tables, the minimum must be used in the above formula. This is because a compression member will buckle about its weak axis – consider the plastic ruler again.

Figure 8.6 shows some buckling curves for steel and aluminium members. It is easy to see how these curves compare with the buckling loads predicted by the Euler formula. We must convert the Euler load to an equivalent stress, σ_E, by dividing by A, the cross-sectional area of the member:

$$\sigma_E = \frac{\pi^2 EI}{AL^2}$$

Now substitute for I from $I = Ar^2$

$$\sigma_E = \frac{\pi^2 EAr^2}{AL^2} = \frac{\pi^2 E}{(L/r)^2}$$

The above Euler equivalent stress can be plotted on the same axes as the other design curves (*figure 8.6*). Do not forget, however, that buckling is principally influenced by stiffness and not strength. The fact that we use stresses to determine compressive strength is largely a convenience. It does not imply that a slender buckled compression member has reached the material yield stress. Note how the yield stress and the Euler equivalent stress form a boundary to the design curves.

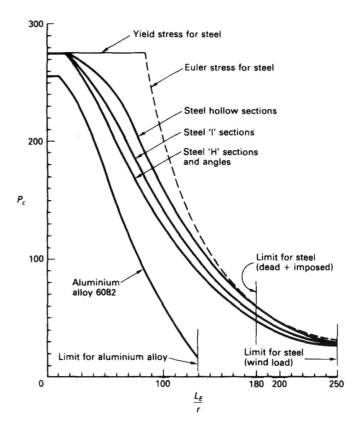

Figure 8.6
Design curves for compression members in
Grade 43 mild steel and aluminium alloy
6082. Based on BS 5950:Part 1: 1990 and
BS 8118:1991

Also note the limits imposed on the maximum permissible slenderness ratios. If these limits are exceeded, the members become too sensitive to small inadvertent errors in straightness, the point of load application and deflections due to self-weight.

Question – If the yield stress for aluminium alloy is almost the same as that for mild steel, why does *figure 8.6* show the buckling load for slender struts in aluminium alloy to be less than half the value for steel?

Answer – For slender struts the buckling load is dependent upon stiffness and not strength. The modulus of elasticity for aluminium alloy is only about a third of that of steel.

The design procedure for real compression members involves a trial-and-error approach as follows:

Step 1   Estimate the compressive strength of the member (say 100 N/mm^2) , and evaluate the approximate area required:

$$\text{Approximate area} = \frac{\text{design load}}{\text{estimated stress}}$$

Step 2   Select a suitable section size from the tables (see *Appendix*) and evaluate the slenderness ratio L_E/r_{min}.

Step 3   Evaluate the compressive strength, p_c, from the design curve, and compare it with the actual stress on the member:

$$\text{Actual stress} = \frac{\text{design load}}{\text{actual area}}$$

Step 4   The actual stress from *step 3* should ideally be less than or equal to the compressive strength obtained from the design curve. If it is higher, then the member will buckle, and the process from *step 2* is repeated using a bigger section. If it is substantially lower, then the process can be repeated with a smaller section, in order to improve economy.

Example 8.3

A pin-ended column in a building is 4 m long and is subjected to the following axial loads:

$$\text{characteristic dead load} = 350 \text{ kN}$$
$$\text{characteristic imposed load} = 300 \text{ kN}$$

Determine the dimensions of a suitable standard universal column section.

Solution

$$\text{Design load} = 1.4G_k + 1.6Q_k$$
$$= 1.4 \times 350 + 1.6 \times 300$$
$$= 970 \text{ kN}$$

Step 1

$$\text{Approximate area} = \frac{970 \times 10^3}{100 \times 10^2} = 97 \text{ cm}^2$$

Step 2

Try 254 × 254 × 89 kg/m universal column
($A = 114$ cm^3, $r_{min} = 6.52$ cm)

Step 3

$$\text{Slenderness ratio} = \frac{L_{\text{E}}}{r_{\text{min}}} = \frac{4000}{6.52 \times 10}$$

$$= 61.3$$

From *figure 8.6*:

Compressive strength, $p_{\text{c}} = 198 \text{ N/mm}^2$

$$\text{Actual stress} = \frac{970 \times 10^3}{114 \times 10^2} = 85 \text{ N/mm}^2$$

Step 4

The above column would be safe but not very economic as the loads produce a stress which is less than half the compressive strength of the member. It is worth trying a smaller section in this case.

Step 2

Try 203 × 203 × 52 kg/m universal column
(A = 66.4 cm³, r_{min} = 5.16 cm)

Step 3

$$\text{Slenderness ratio} = \frac{L_{\text{E}}}{r_{\text{min}}} = \frac{4000}{5.16 \times 10}$$

$$= 77.5$$

From *figure 8.6*:

Compressive strength $p_{\text{c}} = 165 \text{ N/mm}^2$

$$\text{Actual stress} = \frac{970 \times 10^3}{66.4 \times 10^2} = 146 \text{ N/mm}^2$$

As 146 N/mm² is less than 165 N/mm² this is satisfactory.

Answer   Use 203 × 203 × 52 kg/m universal column

Comment – The original guess of a stress of 100 N/mm² was clearly not very good. Experience shows that a value of about 150 N/mm² is appropriate for large columns, whereas a lower value of 100 N/mm² or less is a reasonable starting point for small struts. However, as shown above, it does not really matter if the first guess is some way out.

8.4.3 Compression members without pinned ends

In real structures it is rare for compression members to have pins at the ends. A strut with welded end connections, such as that shown in *figure 8.7a*, will support a larger compressive load before buckling than a pin-ended strut of the same length. This is because the welded connections provide some restraint against rotation. On the other hand a column with a free end, such as that shown in *figure 8.7b*, will buckle at a lower load than a pin-ended column of the same length. The design of such cases is simply taken care of by modifying the effective length before evaluating the slenderness ratio. This is done in accordance with *table 8.1*.

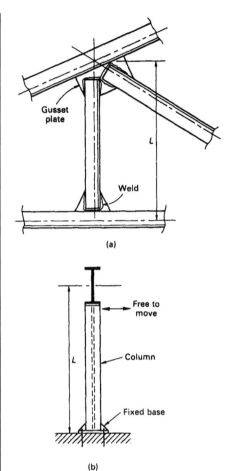

(a)

(b)

Figure 8.7
Examples of compression members that do not have pinned ends

Table 8.1
Effective length, L_E for compression members

End conditions	Effective length, L_E
Both ends fully fixed in position and direction (Must be rigidly fixed to massive members at ends)	$0.7L$
Both ends restrained in position and partially restrained in direction (At least two bolts or welding at each end)	$0.85L$
Fully fixed at one end and free at the other (A vertical cantilever)	$2.0L$

L is the actual length of the member and should be measured from the intersections of the centroids of connecting members.
Based on BS5950: Part 1: 1990

If we apply the values in *table 8.1* to the two cases in *figure 8.7*:

for *case a* $L_E = 0.85L$ (the same would apply if the weld was replaced by two or more bolts)

for *case b* $L_E = 2.0L$

Evaluating effective lengths often requires the use of 'engineering judgement', and often seems rather arbitrary. It certainly makes a nonsense of so-called sophisticated computer analyses that give compressive loads in members to say four decimal places. The estimation of the effective length makes much more difference to the final section size used.

Example 8.4

Figure 8.8 shows a simple crane for lifting loads off vehicles outside a warehouse. It is fixed to a concrete wall. The diagonal bracing shown in plan is to prevent any lateral movement of the point B, and does not affect the forces in the main members.

a. If the dead load of the hoist is 2 kN and the maximum imposed load is 50 kN, determine the forces in the members. (Ignore the self-weight of the structure.)
b. Design a suitable mild steel tie and strut. Use 6 mm thick flat bar for the tie and an equal-angle for the strut.

Solution

Load   Design load $= (1.4 \times 2) + (1.6 \times 50) = 82.8$ kN

Analysis

Resolve forces at joint B – *figure 8.8b* – note 3:4:5 triangle
$(\Sigma V = 0)$ $82.8 = F_{AB} \sin \theta = F_{AB} \times 4/5$

Answer $F_{AB} = 103.5$ **kN tension**

$(\Sigma H = 0)$ F_{BC} $103.5 \times 3/5$

Answer $F_{BC} = 62.1$ **kN compression**

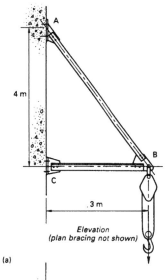

4 m

3 m

Elevation
(plan bracing not shown)

(a)

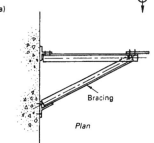

Bracing

Plan

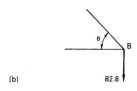

(b)   82.8

Figure 8.8
A simple crane structure

Design

Tie – As in *chapter 6* we take the characteristic strength of mild steel as 275 N/mm^2 and use a partial safety factor for material strength, γ_m of 1.0.

$$\text{Net area required} = \frac{103.5 \times 10^3}{275} = 376 \text{ mm}^2$$

$$\text{Width of flat tie} = \frac{376}{6} + 18 = 81 \text{ mm}$$

Answer   Use 90 mm × 6 mm flat bar

Strut – We shall use the design curve – *figure 8.6*

$$\text{Effective length, } L_E = 0.85 \times 3000 = 2550 \text{ mm}$$

$$\text{Approximate area} = \frac{62.1 \times 10^3}{100 \times 10^2} = 6.2 \text{ cm}^2$$

Try 60 × 60 × 6 angle
(A = 6.91 cm^2, r_{min} = 1.37 cm)

$$\text{Slenderness ratio} = \frac{2550}{13.7} = 186 > 180 \text{ so no good!}$$

Try 80 × 80 × 6 angle
(A = 9.35 cm^2, r_{min} = 1.57 cm)

$$\text{Slenderness ratio} = \frac{2550}{15.7} = 162$$

From *figure 8.6*:

$$\text{Compressive strength } p_c = 60 \text{ N/mm}^2$$

$$\text{Actual stress} = \frac{62.1 \times 10^3}{9.35 \times 10^2} = 66.4 \text{ N/mm}^2$$

66.4 N/mm^2 > 60 N/mm^2 – *still no good!*

Try 90 × 90 × 6 angle
(A = 10.6 cm^2, r_{min} = 1.78 cm)

$$\text{Slenderness ratio} = \frac{2550}{17.8} = 143$$

From *figure 8.6*:

$$\text{Compressive strength } p_c = 74 \text{ N/mm}^2$$

$$\text{Actual stress} = \frac{62.1 \times 10^3}{10.6 \times 10^2} = 58.4 \text{ N/mm}^2$$

$$58.4 \text{ N/mm}^2 < 74 \text{ N/mm}^2 - \text{satisfactory}$$

Answer   Use 90 × 90 × 6 angle

8.5 Summary of key points from chapter 8

1. Some masonry and concrete columns are classed as 'short' and fail by compressive crushing of the material.
2. Most structural compressive members (struts) are slender and fail by sideways buckling. The buckling load is related to the **stiffness** of the member rather than its strength.
3. For a slender pin-ended straight strut the **Euler buckling load** is given by:

$$P_{crit} = \frac{\pi^2\, EI}{L^2}$$

4. Real struts tend to be designed using design charts which take into account imperfections in the member. The strength is based on the **slenderness ratio** of the member:

$$\text{slenderness ratio} = \frac{\text{effective length}}{\text{radius of gyration}}$$

This is often an iterative (trial-and-error) process as the radius of gyration, r, is not known until a trial member has been selected.
5. The effective length of compression members without pin ends must be adjusted before the slenderness ratio is evaluated.

8.6 Exercises

E8.1   A short reinforced concrete column measures 400 mm ×
400 mm and contains four 25 mm diameter high yield reinforcing
bars (f_y = 460 N/mm²). If the concrete characteristic strength is
40 N/mm², determine the ultimate axial design capacity of the
column.
3238 kN

E8.2   A square steel rod has cross-sectional dimensions of 10 mm
× 10 mm. It is used as a pin-ended strut with an effective length of
900 mm. Determine its Euler buckling load. (Use the I value from
figure 7.31.)
2.8 kN

E8.3   A 'goal-post' type frame is used as a support for a hoist
which is used to unload heavy vehicles, see *figure 8.9*. Maximum
characteristic imposed load = 250 kN at mid-span. All other loads
can be ignored. The columns are rigidly connected at the base but
unrestrained at the top. Use the design curve from *figure 8.6* to
determine the dimensions of a suitable circular hollow section
(CHS) for the supporting columns.
Use say 168.3 × 5.0 CHS

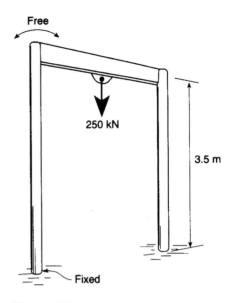

Figure 8.9
'Goal-post' frame for hoist

Chapter 9

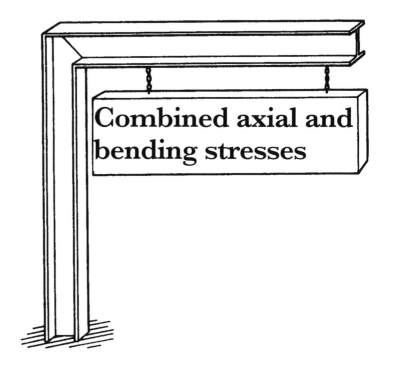

Combined axial and bending stresses

Topics covered

9.1 Introduction

So far we have looked at members subjected to either pure tension, compression or bending. However, in real structures, members are often subjected to a combination of axial force and bending. This chapter is concerned with how the resulting stresses are combined. Consideration is given to both an elastic serviceability approach (i.e. what are the stresses under working loads?) and an ultimate design approach (i.e. is there an adequate safety factor against ultimate collapse?).

We go on to consider **prestressing**, which is the deliberate introduction of beneficial stress into a structure. For example, we can introduce compressive stress into a beam to prevent tensile cracking.

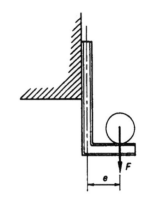

Figure 9.1a
A vertical tension member subjected to eccentric load

9.2 Examples of structures with combined stresses

Wherever a force is **not** applied through the centroidal axis of a member it produces a moment in addition to an axial force. *Figure 9.1a* shows a suspended bracket for supporting a large pipe. It can be seen that the vertical member is subjected to both a tensile force, F, and a bending moment due to the eccentricity, e, of the load.

Figure 9.1b shows a factory portal frame column which has a bracket attached for an overhead crane beam. The reaction force from the crane is applied eccentrically to the axis of the column. The column is therefore in both compression and bending.

The simply supported beam shown in *figure 9.1c* is subject to bending from the applied loads plus compression from the tensioned cable. The fact that the cable force is not applied on the beam centroid means that it also imposes a bending moment on the beam. In fact, as we shall see later, the two bending moments produce opposite stresses which prevent the beam from cracking.

Figure 9.1d shows a brickwork chimney-stack on a house with a wind force acting on the side. The bending moment from the wind force will create tensile stresses on the outside face of the stack, and brickwork has poor strength in tension. However, the self-weight of the brickwork produces compressive stresses which maintain stability.

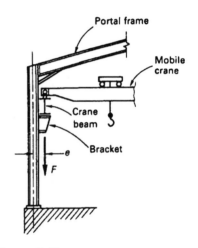

Figure 9.1b
A column with eccentric load from a crane beam

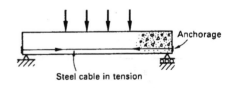

Figure 9.1c
A prestressed concrete beam

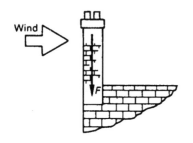

Figure 9.1d
Stability due to self-weight

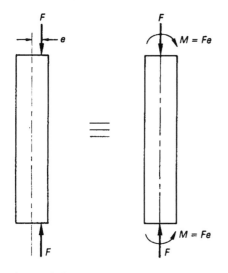

Figure 9.2
The equivalent moment for an eccentric load

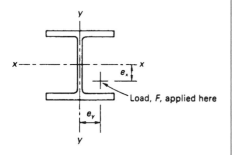

Figure 9.3
A load which is eccentric about two axes

9.3 Eccentrically applied forces

Figure 9.2 shows a force, *P*, applied to a member at some distance, *e*, from the centroidal axis. The distance *e* is known as the **eccentricity**. For design purposes this is exactly equivalent to applying the force at the centroid plus applying a moment equal to the force times the eccentricity.

$$\text{Moment} = F \times e \tag{9.1}$$

In *figure 9.3* a universal column is subjected to a force which is eccentric about both axes. In this case it is necessary to evaluate two moments:

$$M_x = Fe_x$$
$$M_y = Fe_y$$

For simple beam and column construction using end-plane connections, the beam reaction force is assumed to act a distance of 100 mm from the face of the end-plate. This imparts a moment to the column owing to the eccentricity shown in *figure 9.4*. However, if beams are connected to opposite sides of the column, then design is based upon the net moment resulting from the difference between the two moments. This is illustrated in the following example. To obtain the initial guess for the column area, use the total axial load divided by a reduced compressive strength, say 100 N/mm².

Example 9.1

Figure 9.5a shows a 203 × 203 × 46 kg/m universal column which supports four beams. The design reaction force from each beam is given. Determine the axial force, *F*, and the moments, M_x and M_y, for which the column must be designed.

Solution

The design axial load is simply the sum of the loads:

Answer   Axial load, *F* = 70 + 210 + 145 + 75 = 500 kN

To obtain the moments we need to extract the precise dimensions of the column from section tables. These are shown in *figure 9.5b*.

Working in metres:

$$M_x = (210 - 75) \times (0.1016 + 0.1)$$

Answer $M_x = \mathbf{27.2\ kNm}$

$$M_y = (145 - 70) \times (0.0037 + 0.1)$$

Answer $M_y = \mathbf{7.8\ kNm}$

9.4 Ultimate design of columns with bending

Where a member is subjected to an axial force plus one or more moments, a simplified and conservative approach is to apply the following formula:

$$\frac{\text{design axial force}}{\text{axial capacity}} + \frac{\text{design } x\text{–}x \text{ moment}}{x\text{–}x \text{ moment capacity}} + \frac{\text{design } y\text{–}y \text{ moment}}{y\text{–}y \text{ moment capacity}} < 1.0$$

This is an **interaction formula**. It simply states that a large applied axial force will reduce the capacity to support moments and vice versa. In detail it can be written:

$$\frac{F}{Ap_c} + \frac{M_x}{p_b S_x} + \frac{M_y}{\sigma_y Z_y} < 1.0 \qquad [9.2]$$

where

F = design axial force
M_x and M_y = design moments about x–x and y–y axes respectively
A = area of column
p_c = compressive strength (*figure 8.6*)
p_b = bending strength (*figure 7.37*)
S_x = plastic section modulus (x–x axis)
σ_y = yield stress
Z_y = elastic section modulus (y–y axis)

Example 9.2

If the column in *example 9.1* has an effective length of 4000 mm, check that the section size is adequate to support the loads.

Solution

From *example 9.1* we know the following loads:

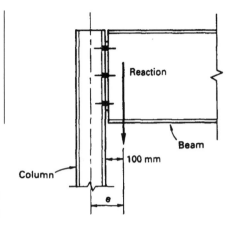

Figure 9.4
Eccentricity in simple beam and column construction

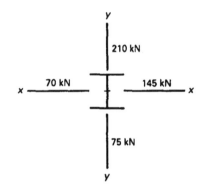

Figure 9.5a
Loads on a column

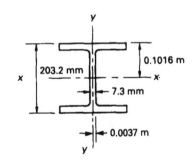

Figure 9.5b
The dimensions of a column from section tables

$$F = 500 \text{ kN}$$
$$M_x = 27.2 \text{ kNm}$$
$$M_y = 7.8 \text{ kNm}$$

We also require additional data from the section tables:

For 203 × 203 × 46 kg/m universal column
($A = 58.8 \text{ cm}^2$, $S_x = 497 \text{ cm}^3$, $Z_y = 151 \text{ cm}^3$, $r_{min} = 5.11 \text{ cm}$, $x = 17.7$)

$$\text{Slenderness ratio} = \frac{L_E}{r_{min}} = \frac{4000}{51.1} = 78$$

from *figure 8.6* compressive strength, $p_c = 161 \text{ N/mm}^2$
from *figure 7.37* bending strength, $p_b = 215 \text{ N/mm}^2$

Therefore
$$Ap_c = 58.8 \times 161 \times 10^{-1} = 947 \text{ kN}$$
$$p_b S_x = 215 \times 497 \times 10^{-3} = 107 \text{ kNm}$$
$$\sigma_y Z_y = 275 \times 151 \times 10^{-3} = 41.5 \text{ kNm}$$

substitute into *equation [9.2]*

$$\frac{500}{947} + \frac{27.2}{107} + \frac{7.8}{41.5} = 0.96 < 1.0$$

Answer   Column is adequate

$$\sigma = \frac{F}{A} \; \mathbb{I} \; \text{▥}$$
Stress distribution

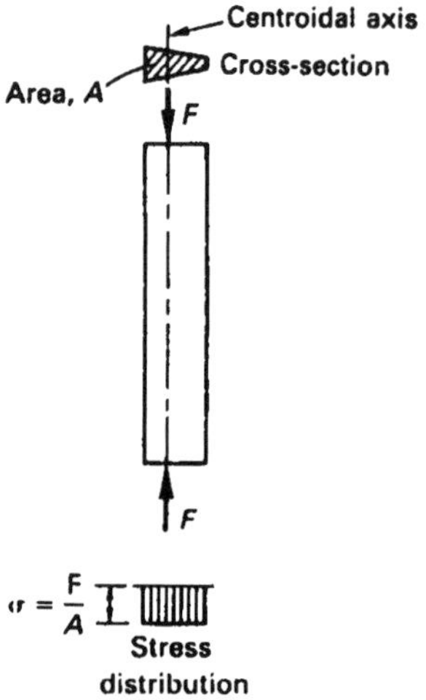

Figure 9.6a
A member with an axial force applied through the centroid

9.5 Combining elastic stresses

We know that if a member supports a tensile or compressive force, F, and it is applied along the centroidal axis, uniform stresses result. This is shown in *figure 9.6a*. The value of the stress, σ, is simply the force, F, divided by the cross-sectional area, A.

$$\sigma = \frac{F}{A}$$

If the same member is subjected to equal and opposite end bending moments, M, as shown in *figure 9.6b*, this will result in a uniform bending moment throughout the length of the member. The resulting stresses will be tensile on one side of the member and compressive on the other. From elastic bending theory in *chapter 7*, the magnitude of the stresses at the surface of the member are given by:

$$\sigma_{max} = \frac{M}{Z}$$

The elastic section modulus, Z, can, of course, have a different value for each side of the member, depending upon the distance from the neutral axis to the surface.

If the member is subjected to **both** an axial force and a moment, the stresses can simply be added together as shown in *figure 9.6c*.

$$\sigma_{max} = \frac{F}{A} + \frac{M}{Z}$$ [9.3]

Equation [9.3] is only true if the stresses remain below the yield point and within the elastic range of the material at all times. Always think carefully about the signs of the axial and bending components before you combine them. The minimum stress is given by:

$$\sigma_{min} = \frac{F}{A} + \frac{M}{Z}$$ [9.4]

Remember that the bending moment can arise from a tension or compression force which is applied eccentrically to the centroid of the member. Moments also arise in bent members, which we shall see in the next example.

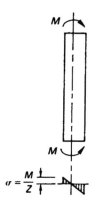

Figure 9.6b
A member with a bending moment applied at the ends

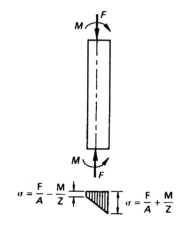

Figure 9.6c
Combined bending and axial force

Example 9.3

Figure 9.7a shows a bent support structure for a ski-lift. It is fabricated from standard circular hollow section of outside diameter 60.3 mm and wall thickness 4.0 mm. Calculate the stresses in the tube at the level *x–x* when supporting two people weighing 100 kg.

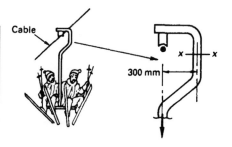

Figure 9.7a
A ski-lift

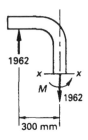

Figure 9.7b
Free-body diagram

Solution

$$\text{Applied force, } F = 200 \times 9.81 = 1962 \text{ N}$$

Refer to the free-body diagram shown in *figure 9.7b*.

$$\text{Moment, } M, \text{ at } x\text{--}x = 1962 \times 300 = 588\,600 \text{ Nmm}$$

From section tables:
For 60.3 × 4.0 circular hollow section
(A = 7.07 cm², Z = 9.34 cm³)

From *equations [9.3]* and *[9.4]*:

$$\sigma = \frac{F}{A} \pm \frac{M}{Z}$$

$$= \frac{1962}{707} \pm \frac{588\,600}{9340} = 2.8 \pm 63$$

**Answer   Stress, σ = 66 N/mm² tension or 60 N/mm²
compression**

9.6 Prestressing

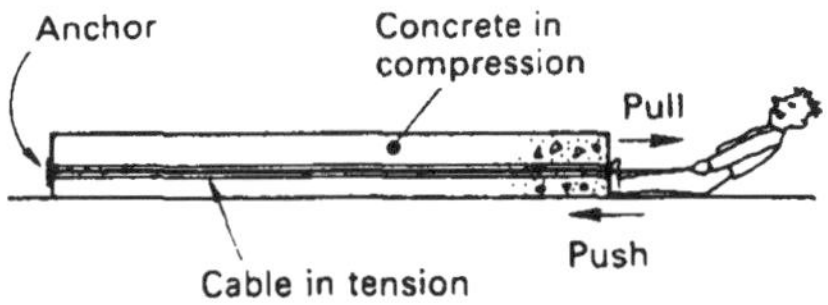

Prestressing is a process in which beneficial stresses are deliberately introduced into a structure. It is usually applied to materials which are strong in compression but relatively weak in tension, such as concrete and sometimes brickwork. In one system highly tensioned cable or tendon is passed through the member and then securely anchored at the ends. The effect is to put the member into compression **before** any external loads are applied. This is illustrated in *figure 9.8*. The advantages are smaller and more efficient structural members, and the elimination of cracks.

There are two basic methods of introducing prestress. The first is **post-tensioned prestressing**. In this approach the concrete is cast in the normal way, with ducts passing through it containing the unstressed prestressing tendons. The concrete is allowed to harden for a sufficient length of time until it is strong enough to take the prestressing forces. The tendons are then carefully stressed with a hydraulic jack which pushes off a steel anchorage at one end of the member. The tendons are locked off by toothed

Figure 9.8
The principle of prestressing

conical wedges which grip the tendon (*figure 9.9*). The duct is then filled under pressure with cement grout, to protect the tendon from corrosion. There has recently been some concern about prestressing tendons corroding through inadequate grouting, and the concept of **external** tendons, which can be easily inspected, is gaining popularity.

The second approach is **pre-tensioned prestressing**. This method is most appropriate for smaller members produced under factory conditions, and then transported to site. The prestressing tendons or single wires are tensioned before the concrete is placed. This requires an external frame capable of resisting the forces. The concrete is placed around the wires and allowed to harden (*figure 9.10*). The wires are then simply cut off at the face of the concrete. Rapid hardening cement and accelerated steam curing are usually used to speed up the turnround of expensive formwork.

Alternatively we could think of an inflatable structure, such as a rubber dingy, as being prestressed. In this case compressed air is used to keep the rubber skin in tension when subjected to bending moments (*figure 9.11*).

Prestressed members in concrete and brickwork can, and should, be checked for ultimate strength. The procedure, for beams, is similar to the one we used in *chapter 7* for the design of reinforced concrete members. However, most of the early stages of prestressed concrete design are concerned with ensuring that the elastic stresses under working loads are kept below acceptable values. In limit state terms this is a serviceability criterion, and consequently the partial safety factors for loads, γ_f, are 1.0. In **class 1 members** there must be no tensile stresses under service loads. In **class 2 members** a small tensile stress of around 3.0 N/mm^2 is permitted.

Figure 9.12a shows a concrete 'T' beam which is subjected to a point load at mid-span. This will produce a bending moment diagram as shown (the self-weight of the beam has been ignored). At the mid-span point the maximum bending moment, M, produces compressive stress in the top of the beam, and tensile stress in the bottom of the beam. The bending stresses at the ends of the beam will be zero since the bending moment is zero.

Now consider the same beam to be completely unloaded but prestressed by a force, F, acting on the neutral axis of the beam. This will produce uniform compression throughout the beam (*figure 9.12b*).

If we now subject the beam to both the point load and the prestressing force, then, provided the stresses remain within the

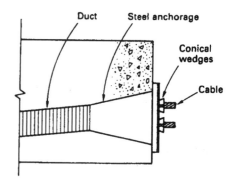

Figure 9.9
A post-tensioned cable anchorage

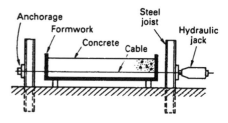

Figure 9.10
A pre-tensioning frame

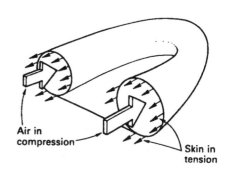

Figure 9.11
A free body of a section of rubber dingy

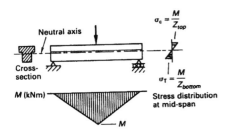

Figure 9.12a

'T' beam with bending

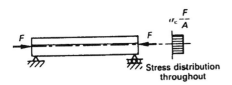

Figure 9.12b

A beam with prestress applied on the neutral axis

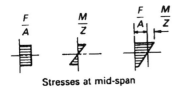

Stresses at mid-span

Figure 9.12c

Combined bending and prestress applied on the neutral axis

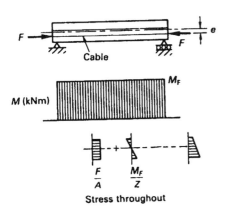

Figure 9.12d

Stresses from eccentric prestress

elastic range, we can apply *equations [9.3]* and *[9.4]* and simply combine the stresses. The result is shown in *figure 9.12c*. Note how, if the prestressing force is big enough, we can completely eliminate tensile stresses from the concrete. The stress at the ends of the beam remains simple uniform compression from the prestress. Although we have eliminated tension from the concrete, we have increased the magnitude of the compressive stresses in the top of the beam. Generally this value must be kept below a third of the 28-day cube strength of the concrete.

We can improve the efficiency of the prestressing process by applying the force, F, with some eccentricity, e, to the beam neutral axis (*figure 9.12d*). It is moved towards the tension face. This produces an additional moment, $M_F = Fe$, throughout the beam which helps to counteract the effect of the applied moment. The prestress produces a hogging moment which is opposite to the applied sagging moment. *Equation [9.4]* can now be extended to:

$$\sigma_{max} = \frac{F}{A} - \frac{M}{Z} + \frac{M_F}{Z} \qquad [9.5]$$

The stresses are shown in *figure 9.12e*. By applying the prestress eccentrically, the required magnitude of the prestressing force is reduced, however, it can lead to a complication. If the eccentricity is too great, the tensile stress from the bending component, M_F, can exceed the compressive stress from F (refer to *figure 9.12d*). This is not a problem at mid-span when the beam is loaded. However, at the ends of the beam (where the applied bending moment is zero), we would get tensile stresses at the top of the beam. To avoid this problem it is sometimes necessary to curve the prestressing cables as shown in *figure 9.12f*. Here the eccentricity is maximum at mid-span, where it is most needed, but zero at the supports.

The initial prestressing force which is applied to a member is subject to significant **losses**. These are caused by:

- relaxation of the steel
- elastic shortening of the concrete
- shrinkage and creep of the concrete
- slippage during anchoring
- friction between the tendon and the duct in post-tensioned beams.

Procedures are available for determining the magnitude of the loss from each of the above, but they often total about 30%.

The procedure for determining the prestressing force, and its eccentricity, to eliminate tensile stresses in a class 1 beam is as follows:

Step 1 – Determine the maximum bending moment on the beam from the applied loads.

Step 2 – Evaluate the resulting maximum tensile stress, σ_t, from $\sigma_t = M/Z$. See *figure 9.13a*.

Step 3 – Decide upon the desired distribution of compressive stress from the prestress. If straight cables are to be used, the best distribution is a triangular one, i.e. zero stress on one face increasing to the value of σ_t obtained in *step 2*. See *figure 9.13b*.

The magnitude of the compressive stress on the neutral axis, σ_{NA}, is determined only by the magnitude of the prestressing force, F. The eccentricity has no effect, because bending effects always produce zero stress on the neutral axis. It follows that $F = \sigma_{NA} \times A$, where A is the cross-sectional area of the beam. If F was applied at the level of the neutral axis, the uniform stress distribution shown in *figure 9.13c* would result. This is clearly not what we want, so we must now determine the eccentricity.

Step 4 – The prestressing force, F, must act below the neutral axis in this case, so as to reduce the compressive stress at the top face and increase it at the bottom face. To determine e, we can use *equation [9.3]* where $M = M_F = Fe$.

$$\sigma_{max} = \frac{F}{A} + \frac{Fe}{Z}$$

σ_{max} is the desired stress at the bottom of the beam. The first term, F/A, is equal to the neutral axis stress, σ_{NA}. The only unknown in the equation is e. The stress distribution from the prestress is now as shown in *figure 9.13d*, which is what is required.

When the stresses resulting from the prestressing force, F, are combined with the stresses from the applied loading we get the distribution shown in *figure 9.13e*.

The above procedure gives the lowest value of F, without producing tensile stresses in the top face of the beam.

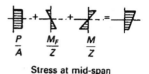

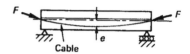

Stress at mid-span

Figure 9.12e
Combined bending and eccentric prestress

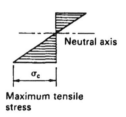

Cable

Figure 9.12f
A beam with a curved prestressing cable

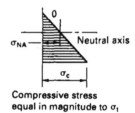

Neutral axis

σ_c

Maximum tensile stress

Figure 9.13a
Stresses from applied loads

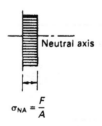

0

σ_{NA} — Neutral axis

σ_c

Compressive stress equal in magnitude to σ_t

Figure 9.13b
The desired stress distribution from the prestress

Neutral axis

$\sigma_{NA} = \dfrac{F}{A}$

Figure 9.13c
The stress distribution if F is applied at the neutral axis

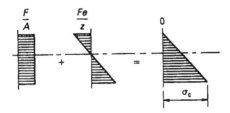

Figure 9.13d
The stress distribution if *F* is applied at an eccentricity of *e*

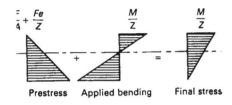

Prestress   Applied bending   Final stress

Figure 9.13e
The tensile stress is eliminated by addition of prestress

Example 9.4

A rectangular concrete beam 1 m deep and 0.5 m wide is simply supported over a span of 10 m. It is subjected to the following uniformly distributed loads:

$$\text{Characteristic dead load} = 20 \text{ kN/m (excluding the self-weight of the beam)}$$
$$\text{Characteristic imposed load} = 24 \text{ kN/m}$$

Determine the minimum prestressing force, and its position relative to the bottom of the beam, if the member is class 1 (i.e. all tensile stresses are to be eliminated). Straight cables are to be used.

Solution

Loads

We need to know the dead load due to the self-weight of the beam. From *table 4.1* the unit weight of concrete is 24 kN/m^3.

$$\text{Dead weight of beam} = 24 \times 1 \times 0.5 = 12 \text{ kN/m}$$

For service loads we use a partial safety factor for loads, γ_f, of value 1.0.

$$\text{Total load on beam} = 12 + 20 + 24 = 56 \text{ kN/m}$$

Analysis

Step 1

$$\text{Max. bending moment} = \frac{wL^2}{8} = \frac{56 \times 10^2}{8} = 700 \text{ kNm}$$

$$\text{Elastic modulus, } Z = \frac{BD^2}{6} = \frac{0.5 \times 1^2}{6} = 0.083 \text{ m}^3$$

Step 2

$$\text{Bending stresses} = \frac{M}{Z} = \frac{700}{0.083} = 8434 \text{ kN/m}^2$$
$$= 8.43 \text{ N/mm}^2$$

The resulting stress distribution is as shown in *figure 9.14a*.

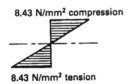

8.43 N/mm² compression

8.43 N/mm² tension

Figure 9.14a
The stresses from applied bending

Design

Step 3

To cancel the tensile stress from the applied loads, the desired stress distribution from the prestress is as shown in *figure 9.14b*.

Stress on neutral axis, $\sigma_{NA} = 8.43/2 = 4.215 \text{ N/mm}^2$

Prestressing force, $F = \sigma_{NA}A$
$$= 4.215 \times 1000 \times 500 \times 10^{-3}$$

Answer   Prestressing force = 2108 kN

Step 4

From
$$\sigma_{max} = \frac{F}{A} + \frac{Fe}{Z}$$

$$8.43 = 4.215 + \frac{2108 \times 10^3 \times e}{0.083 \times 10^9}$$

hence   Eccentricity, $e = 166$ mm
Distance from bottom $= 500 - 166 = 334$ mm

Answer   Distance from bottom to centre of prestress = 334 mm

The final stress distribution at mid-span is shown in *figure 9.14c*.

Figure 9.14b
The desired stress distribution from prestress

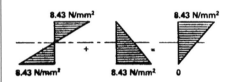

Figure 9.14c
The final stress distribution at mid-span

Under differing load conditions, a particular point on a beam can be subjected to both sagging and hogging moments. This means that tensile stress from the applied moments can occur on either face of the beam. It is possible to optimise the magnitude and position of the prestressing force to cope with tensile stresses on either face. Instead of being triangular in shape, the desired stress distribution, which results from the prestressing force, becomes trapezoidal. This is illustrated in the following example.

Example 9.5

Consider the prestressed beam from *example 9.4*. In addition to the previous load case, the beam is to be lifted into place by a crane which is attached to the beam at third points, and hence causes hogging moments (*figure 9.15a*).

a. Determine the new magnitude and position of the prestressing force if no tensile stresses are to form under either load case.

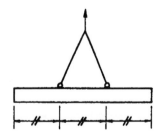

Figure 9.15a
Lifting the beam

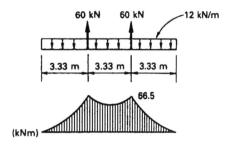

Figure 9.15b
The bending moment diagram from lifting

b. Determine the magnitude of the maximum compressive stress in the beam, and hence specify a minimum concrete grade for the beam.

Solution

Analysis

The first stage is to determine the maximum hogging moment that is caused by the lifting process. The loading consists of the uniformly distributed self-weight of the beam = 12 kN/m. *Figure 9.15b* shows the structural model and the resulting bending moment diagram.

$$M_{max} = 12 \times 3.33 \times 3.33/2 = 66.5 \text{ kNm}$$

$$\text{Bending stresses} = \frac{M}{Z} = \frac{66.5}{0.083} = 801 \text{ kN/m}^2$$

$$= 0.80 \text{ N/mm}^2$$

Design

a.

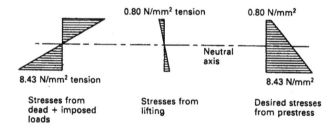

Figure 9.15c
The applied stresses and the desired stresses from the prestress

Figure 9.15c shows the stress distributions resulting from the two applied load cases, plus the desired stresses from the prestressing.

$$\text{Stress on neutral axis, } \sigma_{NA} = \frac{8.43 + 0.80}{2}$$

$$= 4.615 \text{ N/mm}^2$$

$$\text{Prestressing force, } F = \sigma_{NA}A$$
$$= 4.615 \times 1000 \times 500 \times 10^{-3}$$

Answer   Prestressing force = 2308 kN

Step 4

From

$$\sigma_{max} = \frac{F}{A} + \frac{Fe}{Z}$$

$$8.43 = 4.615 + \frac{2308 \times 10^3 \times e}{0.083 \times 10^9}$$

hence   Eccentricity, $e = 137$ mm

Distance from bottom $= 500 - 137 = 363$ mm

Answer   Distance from bottom to centre of prestress = 363 mm

Figure 9.15d shows the stresses resulting from adding the prestressing force to each of the other two load cases. It can be seen that the tensile stresses are eliminated.

b.
By inspection of the stress distribution diagrams shown in *figure 9.15d* it can be seen that both cases result in:

Answer   Maximum compressive stress = 9.23 N/mm²

The minimum 28-day compressive strength of the concrete must be three times the above figure.

Answer   Minimum concrete grade = Grade C30

9.7 Masonry structures with wind loads

When a horizontal wind force blows onto a vertical wall it causes bending. When a wall is built into a steel frame or is braced by perpendicular walls the wind load is supported by both horizontal and vertical spanning. Brickwork is stronger when spanning horizontally because of the interlocking of the bricks. However, a two-way-spanning wall panel is statically indeterminate, and hence its analysis is beyond the scope of this book. Long walls without vertical supports, such as boundary walls (*figure 9.16*), must depend upon only the vertical bending strength of the masonry.

Masonry structures can usually be assumed to support a small tensile stress, although a large portion of their lateral strength comes from the prestressing effect of self-weight. In other words,

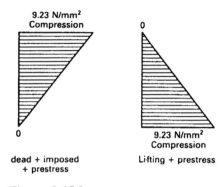

Figure 9.15d
Stresses after adding prestressing force

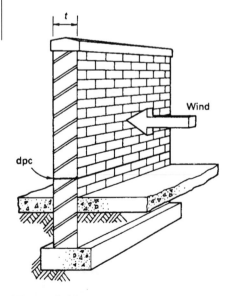

Figure 9.16
A free-standing wall subjected to wind load

brickwork at a particular level in a wall is normally in compression because of the downward weight of the brickwork above that level. Structures which depend upon their own self-weight in this way are known as **gravity structures**. Brickwork is now designed according to limit state principles, and consequently both the loads and material strengths must be subject to partial safety factors. For vertical spanning, the value of the characteristic tensile bending strength, f_{kx}, ranges between 0.2 and 0.7 N/mm^2, depending upon the strength of the mortar and the type and absorbency of the masonry units. A typical value of f_{kx}, is 0.3 N/mm^2.

A situation where masonry has zero tensile strength occurs at most damp proof courses. A damp proof course, or dpc, is an impermeable waterproof layer, usually installed about 150 mm above ground level. The purpose of it is to stop water rising from the ground by capillary action. It usually takes the form of a thin plastic membrane, which cannot transmit tensile stresses. (Special types of impermeable brick are available, and these can transmit tensile stresses.)

Because of these two cases (tension and no-tension) we need to develop two expressions for the maximum design bending moment that a masonry wall can support.

Case 1 – tension case

Where the use of a characteristic tensile bending stress is appropriate, we can use *equation [9.4]* to derive an expression for the maximum design bending moment that a wall can support:

Equation [9.4]
$$\sigma_{min} = \frac{F}{A} - \frac{M}{Z}$$

Because some tension is allowed, the value of σ_{min} in the above equation is allowed to become negative, but must not exceed f_{kx}/γ_m. Therefore at the limiting case:

$$-\frac{f_{kx}}{\gamma_m} = \frac{F}{A} - \frac{M_{max}}{Z}$$

Rearranging
$$M_{max} = (F/A + f_{kx}/\gamma_m)Z \qquad [9.6]$$

where $F =$ design axial load in the wall

Note – When using equation [9.6], table 4.10 implies a partial safety factor on dead load of 1.0. However, unlike the steel and concrete standards,

BS5628 for masonry suggests reducing it to 0.9 when the load is beneficial, as in this case.

Case 2 – no-tension case

If, because of a dpc, the bending tension must be zero, we can use vertical equilibrium to determine the limiting design moment on a wall of thickness t. The characteristic compressive strength of masonry is f_k. We therefore assume that, at the ultimate point of failure, the vertical force in the wall, F, is supported by a strip of material of width x which is at its maximum design compressive stress $= f_k/\gamma_m$ (*figure 9.17*). For a unit length of wall:

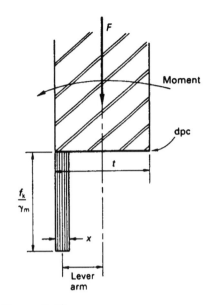

$$(\Sigma V = 0) \qquad F = \frac{f_k}{\gamma_m} \times x$$

$$x = \frac{F\gamma_m}{f_k}$$

$$M_{max} = F \times \text{lever arm}$$

where

$$\text{lever arm} = \frac{t}{2} - \frac{x}{2}$$

Figure 9.17
Stresses at failure for the no-tension case

Therefore substituting for x:

$$M_{max} = \frac{F}{2} \times (t - F\gamma_m/f_k) \qquad [9.7]$$

Example 9.6

A boundary wall is to be constructed from facing bricks and is to be 1.8 m high. It is to have a plastic damp proof course at a height of 150 mm above ground level as shown in *figure 9.18*. What is the required thickness of the wall, based on the following data?

Characteristic tensile bending strength, $f_{kx} = 0.3/\text{mm}^2$
Characteristic compressive strength, $f_k = 7.1 \text{ N/mm}^2$
Characteristic wind load $= 0.5 \text{ kN/m}^2$
Unit weight of brickwork $= 22.0 \text{ kN/m}^3$
Partial safety factor for material strength, $\gamma_m = 3.5$
Partial safety factor on dead load $= 0.9$

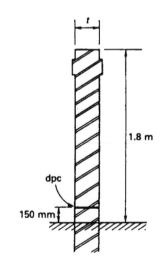

Figure 9.18
A boundary wall with wind load

Solution

Loads

Design wind load $= 1.4 \times 0.5 = 0.7$ kN/m^2
Design dead load $= 0.9 \times 22 \times 0.215 = 4.26$ kN/m^2

Analysis – Consider a 1 m length of wall – for a cantilever:

$$M_{max} = \frac{wH^2}{2} = \frac{0.7 \times 1.8^2}{2} = 1.134 \text{ kNm}$$

also

$$M_{dpc} = \frac{0.7 \times 1.65^2}{2} = 0.953 \text{ kNm}$$

Design – First check the no-tension case at the dpc from *equation [9.7]*:

$$M_{max} = \frac{F}{2} \times (t - F\gamma_m/f_k)$$

where

$$F = 4.26 \times 1.65 = 7.03 \text{ kN}$$

Hence

$$M_{max} = \frac{7.03}{2} \times (t - (7.03 \times 3.5)/7100)$$

$$= 3.52t - 0.0122 \text{ kNm}$$

Equate this with the moment at dpc level from above:

$$0.953 = 3.52t - 0.0122$$

hence

$$t = 0.274 \text{ m}$$

We therefore require the nearest standard brick dimension above this thickness. This is 327.5 mm.

Check the tension case at the base of the wall.

From *equation [9.6]*:

$$M_{max} = (F/A + f_{kx}/\gamma_m)Z$$

where

$$F = 4.26 \times 1.8 = 7.67 \text{ kN}$$

Hence

$$M_{max} = \left\{ \frac{7.67}{0.3275} + \frac{0.3 \times 10^3}{3.5} \right\} \times \frac{0.3275^2}{6}$$

$$= 1.95 \text{ kNm} > 1.134 \text{ kNm}$$

Answer   Use wall 0.3275 mm thick

Comment – In practice it would be more economical to use a thinner wall, but to strengthen it with regular piers, say every two metres throughout its length.

9.8 Summary of key points from chapter 9

1. Wherever a tensile or compressive force is applied eccentric to the centroidal axis of a member, it induces a bending effect. The axial and bending effects must be combined when the member is designed.
2. A conservative approach for the ultimate design of columns with bending is to use an **interaction formula** such as:

$$\frac{F}{Ap_c} + \frac{M_x}{p_b S_x} + \frac{M_y}{\sigma_y Z_y} < 1.0$$

3. Elastic stresses can be combined by adding the axial stresses to the bending stresses:

$$\sigma = \frac{F}{A} \pm \frac{M}{Z}$$

4. Prestressing is a process where an additional compressive force is applied to a member in such a way that unwanted tensile stresses from bending, M, can be reduced or removed. The force, F, is usually applied with an eccentricity, e, to the beam centroidal axis so that it counteracts the bending effect. The above equation becomes:

$$\sigma = \frac{F}{A} \pm \frac{M}{Z} \mp \frac{Fe}{Z}$$

5. Masonry structures often rely on their own dead-weight to provide a compressive pre-stress in order to resist tensile stresses from bending.

9.9 Exercises

E9.1   A column with an effective length of 6.0 m is subjected to the design loads shown in *figure 9.19* from three beams. Determine the dimensions of a suitable universal column.
Use 305 × 305 × 137 UC

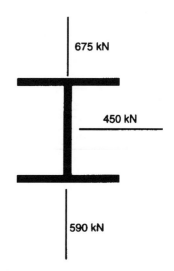

Figure 9.19
Column supporting three beams

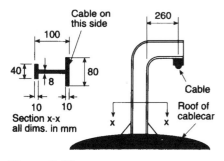

Cable on this side

Section x-x
all dims. in mm

Figure 9.20

E9.2   *Figure 9.20* shows the support for a cable-car which can carry up to thirty people (average mass = 70 kg). It has a self-weight of 2000 kg. It is supported by a single steel member with a cross-section as shown in x–x.

a) Calculate the elastic section moduli for the cross-section.
b) Determine the maximum elastic tensile and compressive stresses in the support member at section x–x when fully loaded. (Do not apply any partial safety factors to the loads.)

a) $Z_{inner} = 64.7 \ cm^3$, $Z_{outer} = 43.5 \ cm^3$
b) $\sigma_{inner} = 208.2 \ N/mm^2$ *tension,* $\sigma_{outer} = 254.2 \ N/mm^2$ *comp.*

E9.3   A particular point on a beam may be subjected to a sagging moment of 24 kNm or a hogging moment of 9 kNm. The beam has the following properties:

$$\text{cross-sectional area, } A = 200 \ cm^2$$
$$\text{second moment of area, } I = 18 \ 000 \ cm^4$$
$$\text{overall depth} = 25 \ cm$$
$$\text{distance from top to centroid} = 10 \ cm$$

What is the least prestressing force and at what depth below the centroid must it be applied if there is to be no tensile stress in the beam?
force = 220 kN, depth below centroid = 49.2 mm

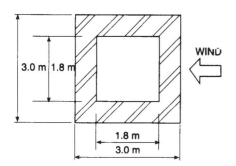

Figure 9.21

E9.4   *Figure 9.21* shows a cross-sectional plan of a brickwork chimney. This cross-section remains constant for the whole height. The chimney is subjected to wind load acting perpendicular to one face. Given:

$$\text{height of chimney} = 30 \ m$$
$$\text{wind pressure varies linearly} = \text{zero at ground level}$$
$$\text{and} = 1.2 \ kN/m^2 \text{ at the top}$$
$$\text{weight of brickwork} = kN/m^2$$

Determine the distribution of stresses over the cross-section at the base of the chimney.
Stress varies from 0.876 N/mm² compression on leeward face to 0.324 N/mm² compression on the windward face

Chapter 10

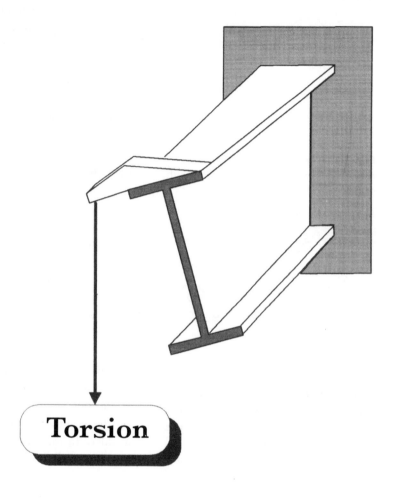

Torsion

Topics covered

The significance of torsion
Solid sections
Closed hollow sections
Open thin sections
Combined bending and torsion

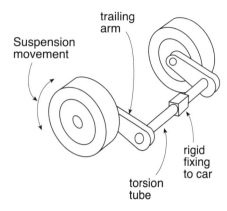

Figure 10.1a
A vehicle trailing-arm suspension with torsion tube

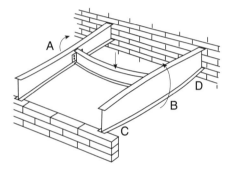

Figure 10.1b
Compatibility torsion. Edge beam forced to rotate at midspan to accommodate sagging of central beam

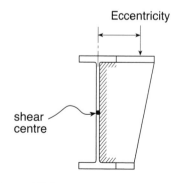

Figure 10.1c
Torsion caused by load offset from shear centre

10.1 Introduction

In earlier chapters we have considered the stresses in a structural member when we pull it, push it or bend it. In this chapter we will consider what happens when we **twist** it. A member twists when it is subjected to a torsional moment or **torque** which like a bending moment has the units of force × distance (kNm).

It is important to distinguish between two types of torsion:

- **Compatibility torsion** arises when a member has to twist in order to accommodate the deflection of another member. *Figure 10.1b* shows a beam arrangement where it can be seen that any sagging deflection of beam A–B will cause some twisting and hence torsion in beam C–D. However, the angle of twist and hence the resulting torsion stresses are limited by the stiffness of beam A–B. Even if beam C–D had end connections that were incapable of resisting any rotational torques, the structure would not collapse. Provided beam A–B is designed as simply supported (i.e. no end fixing moments assumed), the **torsion in beam C–D can be ignored**. This is generally the case with compatibility torsion.

- **Equilibrium torsion** is much more significant. For the case shown in *figure 10.1c* both the structural member and its end connections must be capable of resisting the torque imposed by the eccentricity of the applied load if collapse is to be avoided. In this case the stresses that result from torsion must be added to those from the normal bending of the beam.

In practice, the best approach to torsion in structural design is to eliminate it where possible. This means, for example with standard 'I' beams, that the beam should be placed centrally under the load where possible.

10.2 Examples of torsion

Torsion is widely used in mechanical engineering as a means of transmitting power from, say, the engine to the wheels of a car. This is usually by means of either a solid or hollow circular drive shaft. A torsion tube also forms an economical means of achieving an independent wheel suspension system (*figure 10.1a*).

Structural members sometimes need to twist in order to accommodate the deflection of another part of the structure (*figure 10.1b*). This is **compatibility torsion** and can usually be ignored provided torsional strength is not required for the maintenance of equilibrium. This is discussed in more detail above.

Torsion occurs whenever the line of action of a load is applied offset to a point on a member cross-section known as the **shear centre**. For a symmetrical beam this coincides with the centroid. *Figure 10.1c* shows a beam supporting a side cantilever which clearly causes twisting and hence torsion in the main beam.

The edge beam shown in *figure 10.1d* supports a load from an external wall that acts eccentrically to the shear centre and hence causes torsion. For non-symmetrical beams the shear centre does not coincide with the centroid. *Figure 10.1e* shows the situation with a standard channel section, where, because the load passes through the shear centre, the beam is not subjected to torsion.

Eccentric horizontal loads can also cause torsion. *Figure 10.1f* shows a crane gantry girder that is subjected to a horizontal load arising from surging movement of the crane in addition to vertical load. This will clearly cause the beam to twist, however it is common practice to ignore torsion effects in this case provided the combined strength of the top flange members are adequate to resist the horizontal bending in addition to the effects of vertical bending.

Figure 10.1g shows a steel box-girder bridge with a concrete deck. If traffic loading acts on only one side of the bridge, the steel box must resist torsion.

10.3 Solid circular shafts

The simplest case to consider is that of a solid circular shaft subjected to a pure torque, *T*, as shown in *figure 10.2a*. If an unloaded shaft was marked with a grid consisting of longitudinal and circumferential lines (*figure 10.2b*), then, after the application of the torque, the longitudinal lines are displaced as shown in *figure 10.2c*. If we consider one rectangle of the grid before and after the application of the torque (*figure 10.2d*) we can see that it has been subjected to shear distortion. Thus in general it can be concluded that **torsion produces shear stress**.

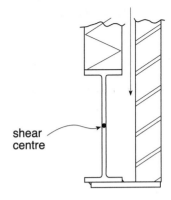

Figure 10.1d
Edge beam subjected to torsion from wall

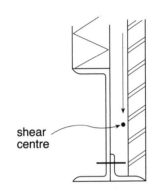

Figure 10.1e
Channel edge beam not subjected to torsion as centre of gravity of load passes through shear centre

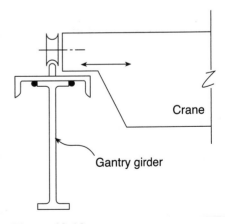

Figure 10.1f
Horizontal crane load causing torsion

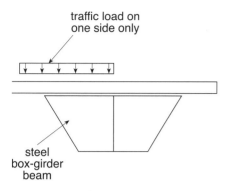

Figure 10.1g
A box-girder bridge with eccentric loading

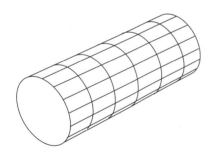

Figure 10.2a
Circular shaft under pure torsion

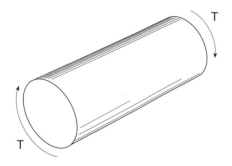

Figure 10.2b
Unloaded shaft with grid

10.3.1 *Ultimate failure of circular shafts*

If the torque is increased until collapse, the failure mode depends upon the nature of the material. A ductile material, such as mild steel, will fail when all of the material on a particular cross-section has reached its shear yield stress τ_y (*figure 10.3a*). The symbol τ is the Greek lower-case letter **tau**. A brittle material, on the other hand, such as cast iron, concrete or glass, will fail in tension producing a characteristic spiral failure surface (*figure 10.3b*).

For the ductile/fully plastic case we can derive the relationship between the applied torque, T, and the shear yield stress, τ_y as follows:

Consider a circular elemental ring of the section as shown in *figure 10.4*. The torque resisted by this single ring at radius, r, is given by:

$$d\tau = \tau_y \times 2\pi r \times dr \times r$$

Integrating over the whole shaft we get:

$$T_{ult} = \tau_y \int_0^R 2\pi r^2 dr$$

$$= \tau_y \left[\frac{2\pi r^3}{3} \right]_0^R$$

$$\boldsymbol{T_{ult} = \frac{2\tau_y \pi R^3}{3}}$$ [10.1]

Example 10.1

Determine the required diameter of a circular mild steel shaft, (τ_y = 165 N/mm^2) if it is to resist an ultimate design torque of 20 kNm.

Solution

$$T_{ult} = \frac{2\tau_y \pi R^3}{3}$$

Rearranging

$$R^3 = \frac{3 T_{ult}}{2\tau_y \pi}$$

$$= \frac{3 \times 20 \times 10^6}{2 \times 165 \times \pi} = 57\ 870 \text{ mm}^3$$

$$R = 38.7 \text{ mm}$$

Answer   **Diameter = 77.4 mm**

10.3.2 *Elastic behaviour of circular shafts*

If the shear stresses on a shaft are kept below the shear yield stress for the material then it is necessary to carry out an elastic analysis. Instead of assuming a uniform stress across the whole cross-section we now assume that the shear stress, τ, is directly proportional to the shear strain, γ. The symbol γ is the Greek lower-case letter **gamma**. For a solid circular shaft this strain is zero at the centre and increases linearly as the radius is increased (*figure 10.5*). The maximum value of shear strain, γ_{max}, and hence shear stress, τ_{max}, thus occurs on the surface of the shaft. (The centre point is clearly analogous to the neutral axis in elastic bending.)

For the elastic case we can derive the relationship between the applied torque, T, and the maximum elastic shear stress, τ_{max}, as follows:

Consider a circular elemental ring of the section as shown in *figure 10.5*. The torque resisted by this single ring at radius, r, is given by:

$$d\tau = \tau \times 2\pi r \times r \times dr$$

where

$$\tau = \tau_{max} \times \frac{r}{R}$$

hence

$$dt = \frac{2\pi\tau_{max}r^3 dr}{R}$$

Integrating over the whole shaft we get:

$$T = \frac{\tau_{max}}{R} \int_0^R 2\pi r^3 dr$$

$$= \frac{\tau_{max}}{R} \left[\frac{2\pi r^4}{3} \right]_0^R$$

$$\boldsymbol{T = \frac{\tau_{max}\pi R^3}{2}}$$ [10.2]

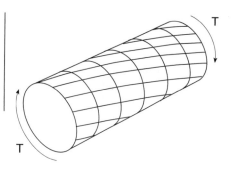

Figure 10.2c
Grid pattern after loading

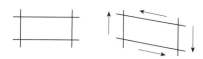

Figure 10.2d
A single rectangle before and after loading

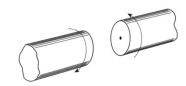

Figure 10.3a
Shear torsion failure in ductile material

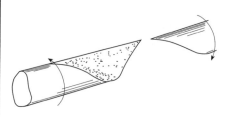

Figure 10.3b
Tensile torsion failure in brittle material

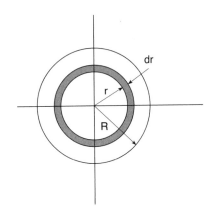

Figure 10.4
An elemental ring subjected to the shear yield stress

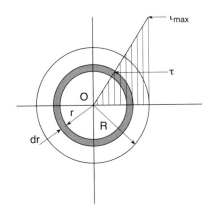

Figure 10.5
A circular shaft under elastic stress

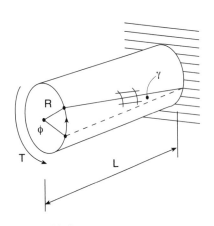

Figure 10.6
Angle of rotation of shaft

10.3.3 *Angle of twist under elastic torsion*

It is an easy matter to calculate the rotational deflection between the two ends of a circular shaft when subjected to torsion. So far we have seen that the relationship between stress and strain is defined by the modulus of elasticity, E:

$$E = \frac{\text{Stress, } \sigma}{\text{Strain, } \varepsilon}$$

In the case of shear the equivalent relationship involves the **shear modulus** or **modulus of rigidity,** G:

$$G = \frac{\text{Shear stress, } \tau}{\text{Shear strain, } \gamma}$$

Hence

$$\gamma = \frac{\tau}{G}$$

The numerical value of G lies somewhere between a half and a third of the value of the modulus of elasticity. It is related to the modulus of elasticity, E, and Poisson's ratio, υ, by the equation:

$$G = \frac{E}{2(1 + \upsilon)}$$

For structural steel G is typically 77 000 N/mm². The shear strain, γ, is defined as the angle shown in *figure 10.6* in radians. If the length of the shaft which is subjected to the torque is L, then the total displacement on the surface of the shaft is simply $L\gamma$. This can then be converted into an angle of rotation of the shaft, ϕ. The symbol ϕ is the Greek lower-case letter **phi**.

$$\phi = \frac{L\gamma}{R}$$

$$\phi = \frac{L\tau}{GR} \qquad [10.3]$$

Example 10.2

A 50 mm diameter circular mild steel shaft, is 1.25 m long and is subjected to a torque of 3 kNm.

Determine:
a) The maximum shear stress in the shaft.
b) The relative angle of twist between the two ends of the shaft.
   $G = 77\ 000\ \text{N/mm}^2$

Solution

a)

from *[10.2]*
$$T = \frac{\tau_{max}\pi R^3}{2}$$

Rearranging
$$\tau_{max} = \frac{2T}{\pi R^3}$$

$$= \frac{2 \times 3 \times 10^6}{\pi \times 25^3}$$

Answer $\boldsymbol{\tau_{max} = 122\ \text{N/mm}^2}$

From *[10.3]*
$$\phi = \frac{L\tau}{GR}$$

$$= \frac{1250 \times 122}{77\ 000 \times 25}$$

$$= 0.0792\ \text{rad}$$

Answer   angle of twist $= \dfrac{0.0792 \times 360}{2\pi} = \boldsymbol{4.5°}$

10.4 Hollow closed sections

Hollow closed sections are the most efficient means of resisting torsion, both in terms of strength and stiffness. For the type of members used in construction, i.e. structural hollow sections, they

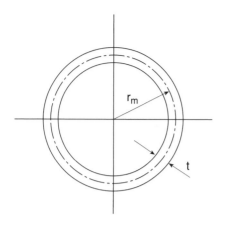

Figure 10.7a
Thin circular hollow section

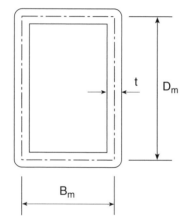

Figure 10.7b
Thin rectangular hollow section

can be regarded as 'thin walled' which simplifies the calculations. Two assumptions are made:

1. The shear stress is assumed constant throughout the wall thickness, t, even for the elastic case.
2. The whole of the material is considered to be concentrated at the centre-line of the wall at radius r_m.

It can be seen from *figure 10.7a* that for a **circular hollow section:**

$$T = \tau \times 2\pi r_m \times t \times r_m$$
$$T = 2\tau\pi r_m^2 t \qquad [10.4]$$

Similarly we could apply the same reasoning to a **rectangular hollow section** (*figure 10.7b*). Simply sum the torque contribution for each side of the section by taking moments about the centre point:

$$T = \tau \times 2[(B_m \times t \times D_m/2) + (D_m \times t \times B_m/2)$$
$$= 2\tau B_m D_m t \qquad [10.5]$$

It can be seen that from both *[10.4]* and *[10.5]* that we could write:

$$T = 2\tau A_m t \qquad [10.6]$$

Where A_m is the area bounded by the centre-line of the wall cross-section.

In fact this is true for all closed thin-walled sections and so *[10.6]* is a general formula that can be applied to any shape. If, in the above equations, τ takes the value of the shear yield stress τ_y, then the equations give the value of the ultimate failure torque, T_{ult}.

10.5 Open thin sections

The underlying theory behind torsion in open thin sections is complex and beyond the scope of this book, however, it is possible to gain some insight into their behaviour from a descriptive treatment. Universal beams, columns and channels are all classed as open sections.

Open sections, such as that shown in *figure 10.8a*, respond to torsion in two ways:

1. Each of the individual flat plates that forms the section twists perpendicular to its own plane (*figure 10.8b*). This is known as **pure torsion**. It produces shear stresses that vary linearly through the thickness of the plate. They are of opposite sign on each side of the plate, and provide relatively little torsional restraint.
2. Longitudinal displacements occur which are evident at the ends of the beam (*figure 10.8c*). This is known as **warping**. In most structures this warping effect is at least partially restrained by the end connections, and this induces longitudinal stresses and shear stresses. These stresses must be added to those from normal bending and pure torsion.

The degree to which torsion is resisted by either of the above mechanisms partly depends upon the amount of warping restraint. Consider the 'I' beam cantilever shown in *figure 10.9* and subjected to a torque at the free end. Provided the beam flanges are rigidly prevented from horizontal rotation at the built-in end, the torque is at least partially resisted by each flange acting as a horizontal cantilever. The top flange is pushed one way, δ_t and the bottom flange the other, δ_b. This action is known as **bi-moments**.

For a thin open section with no warping restraint and constant material thickness, t, the maximum shear stress can be approximated from the following formula:

$$\tau_{max} = \frac{3T}{at^2} \qquad [10.7]$$

where
a = length of plate that makes up the cross-section – see *figure 10.10*

and angle of twist, $\phi = \dfrac{3TL}{at^3G} \qquad [10.8]$

The actual shape of the section has virtually no influence.

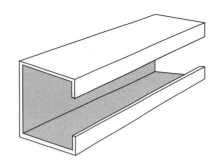

Figure 10.8a
An open thin-walled section

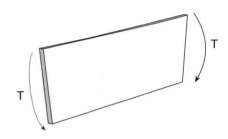

Figure 10.8b
Each plate in thin walled sections is subjected to pure torsion

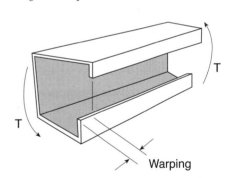

Figure 10.8c
Longitudinal displacements in open sections are known as warping

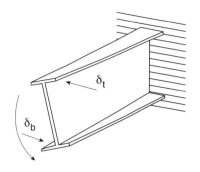

Figure 10.9
Warping restraint at the fixed end causes top and bottom flanges to act as horizontal cantilevers

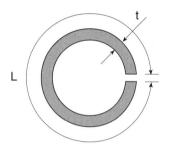

Figure 10.10
A circular hollow section which has been slit to form a thin open section

Example 10.3

A structural member 2.5 m long is required to resist a design torque of 55 Nm. Evaluate and compare both the maximum shear stress and angle of twist for the following trial sections:

a) A 60.3 × 5.0 circular hollow section.
b) A 60.3 × 5.0 circular hollow which has been sawn longitudinally so as to form an open section (*figure 10.10*).
   $G = 77\,000$ N/mm^2

Solution

a) closed section

$$r_m = 30.15 - 2.5 = 27.65 \text{ mm}$$

From *[10.4]* $T = 2\tau\pi r_m^2 t$

$$\tau = \frac{T}{2\tau\pi r_m^2 t}$$

$$= \frac{55 \times 10^3}{2\pi \times 27.65^2 \times 5}$$

Answer   shear stress, $\tau = 4.6$ N/mm^2

From *[10.3]* $\phi = \dfrac{L\tau}{GR}$

$$= \frac{2500 \times 4.6}{77\,000 \times 30.15}$$

Answer   angle of twist, $\phi = 0.0049$ rad   (0.28°)

b) open section

$$a = 2 \times \pi \times 27.65 = 173.7$$

From *[10.7]* $\tau_{max} = \dfrac{3T}{at^2}$

$$= \frac{3 \times 55 \times 10^3}{173.5 \times 5^2}$$

Answer   shear stress, $\tau_{max} = 38$ N/mm^2

From [10.8] $\phi = \dfrac{3\,TL}{at^3G}$

$$= \frac{3 \times 55 \times 10^3 \times 2500}{173.7 \times 5^3 \times 77\,000}$$

Answer   **angle of twist, $\phi = 0.247$ rad**   (14.1°)

Note
It can be seen from the above that the torque generates eight times the stress in the open form of the section compared to the closed form. The twist angle is fifty times as great! It can be concluded that:

1. *Where high torsional stiffness is desirable, such as with car bodies, it is best to use closed box sections.*
2. *Where a given angle of twist is imposed on a member, such as in compatibility torsion, the additional flexibility of an open section will result in lower stresses.*

10.6 Combined bending and torsion

In many structures, such as beams with eccentric loads, stresses resulting from torsion must be combined with those resulting from bending. In the case of open sections with warping restraint this can be quite complex as both forms of loading produce both shear and longitudinal stresses. It may be necessary to combine these stresses at several points in a beam in order to check that the sum does not exceed the yield value.

We will consider the simpler case of a closed hollow section beam subject to torsion. With closed sections torsion produces mainly shear stresses.

Example 10.4

A 3.0 m long beam supports a design load of 60 kN which acts through a short cantilever 0.5 m from one end. The load acts with an eccentricity of 0.1 m from the centre-line of the beam as shown in *figure 10.11a*. The end connections can be assumed to provide torsional restraint but to act as simple supports for the purposes of bending.

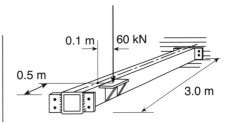

Figure 10.11a
Rectangular hollow section beam with torsion

a) Draw diagrams indicating the distribution of bending moment, shear force and torque.

b) Determine the dimensions of a suitable square hollow section. $\sigma_y = 275$ N/mm^2, $\tau_y = 165$ N/mm^2

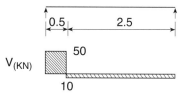

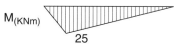

V$_{(KN)}$

Figure 10.11b
Shear force and bending moment diagrams

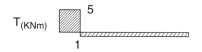

T$_{(KNm)}$

Figure 10.11c
Torque diagram

Solution

a)

$\sum M$ about B $R_A = 60 \times \dfrac{2.5}{3} = 50$ kN

$\sum V = 0$ $R_B = 60 - 50 = 10$ kN

$M_{max} = 50 \times 0.5 = 25$ kNm

Answer – For shear force and bending moment diagrams see *figure 10.11b*

Applied torque, $T = 60 \times 0.1 = 6$ kNm

Note
In order to support the torque it is only necessary for one end of the beam to be able to resist torsion. In this case both ends can do so, therefore in terms of torsion this structure is indeterminate. Both ends of the beam will resist torque, however, it will divide between the two ends in proportion to the torsional stiffness of the beam at each side of the load. This is clearly in inverse proportion to the beam lengths. The result, in this case, is a torque distribution similar to the shear force (figure 10.11c)

b)
The approach is to determine the initial dimensions of the beam based on bending and then to check the combined shear stress resulting from bending and torsion.

$$\text{Required plastic modulus, } S = \frac{25 \times 10^6}{275 \times 10^3}$$

$$= 90.9 \text{ cm}^3$$

From *Appendix*:
*Try 100 × 100 × 8 square hollow section
(S = 99.9 cm^3)*

The shear stress from bending is assumed to be the average value on the web as shown in *section 7.6*.

$$\tau_{bending} = \frac{50 \times 10^3}{2 \times 100 \times 8} = 31.3 \text{ N/mm}^2$$

The shear stress from torsion is given by *equation [10.6]*

$$T = 2\tau A_m t$$

Where

$$A_m = 92 \times 92 = 8464 \text{ mm}^2$$

$$\tau_{torsion} = \frac{T}{2A_m t} = \frac{5 \times 10^6}{2 \times 8464 \times 8}$$

$$= 36.9 \text{ N/mm}^2$$

The above torsional shear stress is added to the bending shear stress at one side of the section and subtracted from the other side. Hence the maximum value is given by:

$$\tau_{total} = 31.3 + 36.9 = 68.2 \text{ N/mm}^2$$
$$(< 165 \text{ N/mm}^2)$$

Answer   Use 100 × 100 × 8 square hollow section

10.7 Summary of key points from chapter 15

1. **Compatibility torsion** occurs when a member twists in order to accommodate the deflection of another member and can usually be ignored. The member could still carry the load even it had no torsional resistance.
2. **Equilibrium torsion** must be resisted if collapse is to be avoided. This requires both the member and its end connection to be able to resist torsion.
3. When a torque is applied to a **solid circular shaft** it produces shear stresses. Simple formulae have been derived to obtain the relationship between the applied torsional load and the elastic shear stresses in the shaft, as well as its angle of twist. It is also simple to calculate the ultimate failure torque that a circular shaft made of plastic material can support.
4. A **hollow closed section** is the most efficient means of supporting a torque, both in terms of strength and stiffness. A simple formula is available to evaluate the stresses in sections where the wall thickness is thin.

5. The theory behind **thin open sections** in torsion is more complex. Such a member is subjected to warping and this produces both longitudinal and shear stresses.
6. When a member is subjected to both **bending and torsion**, the resulting stresses must be combined.

10.8 Exercises

E10.1   A solid circular mild steel shaft has a diameter of 30 mm and $\tau_y = 165$ N/mm^2.

a) What ultimate design torque will it resist?
b) If the shaft is 3 m long and the stress is limited to 90 N/mm^2, determine the relative angle of twist between the two ends. ($G = 77\ 000$ N/mm^2).

a) *1.16 kNm*
b) *0.234 rad (13.4°)*

E10.2   *Figure 10.12* shows a traffic message display sign which is securely fixed to a wall by a cantilever beam. It is subjected to a design horizontal wind load of 1.1 kN/m^2. Determine:

a) The horizontal bending moment and torque on the cantilever support.
b) The dimensions of a suitable square hollow section. Assume that vertical loads are negligible.
    $\sigma_y = 275$ N/mm^2, $\tau_y = 165$ N/mm^2
    Hint: The general approach is as in example 10.4.
a) *Bending moment = 4.98 kNm, torque = 1.51 kNm*
b) *60 × 60 × 4.0 square hollow section ($\tau_{max} = 63.9$ N/mm^2)*

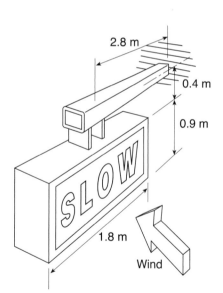

Figure 10.12
Traffic message display with wind load

E10.3   A 150 × 75 × 10 unequal angle is 4 m long and is subjected to a pure design torque of 1.2 kNm at one end. Assuming no warping restraint, determine:

a) The resulting shear stress.
b) Angle of twist ($G = 77\ 000$ N/mm^2).
a) *160 N/mm^2*
b) *0.831 rad (47.6°)*

Chapter 11

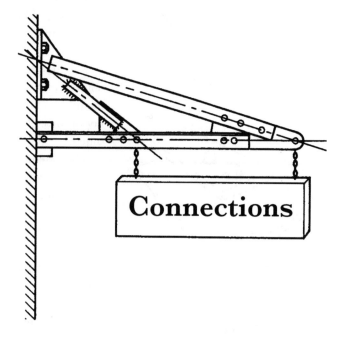

Connections

Topics covered

Examples of connectors and connection details
Welded connections

Bolted connections:
   Ordinary bolts, Friction grip bolts
   Bolt spacing

Beam splices
Connections in built-up beams

Figure 11.1a
A fillet weld between two plates

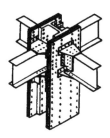

Figure 11.1b
A bolted connection

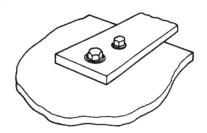

Figure 11.1c
A riveted connection in the Empire State Building, New York, USA

Figure 11.1d
A glued laminated timber beam

11.1 Introduction

In previous chapters we have learned how to analyse the forces in structures, and how to design members in tension, compression and bending. In this chapter we shall consider how individual members are connected together to form complete structures.

Research has shown that the second largest cause of structural failures (30%) is defective detail design of the joints between members. (The largest cause is collapse during construction.) The importance of sound connection design cannot therefore be overstated.

This chapter starts by considering the principal types of connector, and then goes on to consider some common connection details.

11.2 Examples of connectors and connection details

Figure 11.1a shows a typical welded connection. Welds such as this, which are roughly triangular in cross-section, are known as **fillet welds**. Structural welding should only be carried out by qualified welders. It is the most economic method of joining steel components in the fabrication shop, but should only be used with caution on construction sites. Aluminium alloys can also be successfully welded, but as already pointed out, there is a significant loss of strength around the weld.

Figure 11.1b shows the equivalent connection made with bolts. Bolts are preferable for site connections. There are several different types of bolt, and these will be considered later in this chapter.

Rivets were once a popular form of connector in structures but are now rarely used (*figure 11.1c*). Many early New York skyscrapers such as the Empire State Building have riveted steel frames, and they can still be seen on many early railway bridges. They are still widely used in the aircraft industry for fixing the aluminium alloy skin to the wings and fuselage, as they are particularly vibration resistant.

The timber beam shown in *figure 11.1d* is built-up from laminates of timber which are glued together. Glues such as epoxy resins are increasingly used for structural connections, but further research is required to fully investigate the influence of temperature, humidity and surface preparation on strength and durability.

Figure 11.1e shows a further range of connectors available for timber structures. These include nails, screws and 'gangnail' plates. The design strengths of these fixings are given in the relevant national standards.

A typical connection in a pin-jointed frame is shown in *figure 11.1f*. It is important to note that the centroidal axes of all the members intersect at a single point. If this is not done, as shown in *figure 11.1g*, the connectors and the structural members must be subjected to bending moments so that moment equilibrium is satisfied. This will produce premature buckling of compression members.

A modern and economical method of connecting beams and columns is by means of welded end-plates (*figure 11.1h*). All the welding is completed in the fabrication shop. The components are then delivered to site for erection by bolting. Note the notch in the top flange of one beam to ensure that the top faces of the flanges are flush after fixing.

Figure 11.1i shows a typical join in a beam. This is known as a **splice**. It is good practice to ensure that the splices do not occur at points of maximum bending moment. Columns can be spliced in a similar manner.

Standard universal beams are available up to 914 mm deep, but beyond that size beams must be made from individual plates. These are known as **plate girders** or **box girders** (*figure 11.1j*). The welded connections between the individual plates must be carefully designed to transmit shear forces, and this is dealt with later in this chapter.

Monolithic joints in reinforced concrete are formed by making the steel reinforcement continuous across the concrete construction joints. *Figure 11.1k* shows a typical junction between a column and a foundation. There are many rules that must be applied for successful reinforced concrete detailing, and they are beyond the scope of this book.

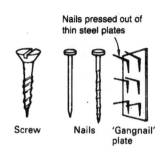

Figure 11.1e
Various timber connectors

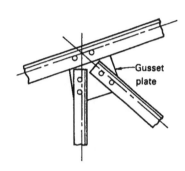

Figure 11.1f
A good truss connection with the centroidal axes of all members meeting at one point

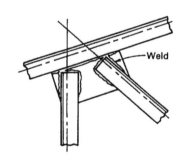

Figure 11.1g
A bad welded connection with the centroidal axes not meeting at a single point

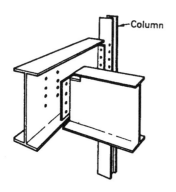

Figure 11.1h
An end-plate connection to a beam and a column

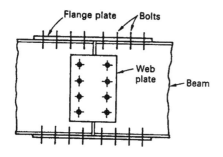

Figure 11.1i
A beam splice

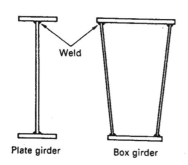

Figure 11.1j
Built-up beams fabricated from welded steel plates

11.3 Welded connections

Most structural welding is now done by the **electric-arc** method, where a welding rod or **electrode** is fused to the parent metal by means of the heat generated by high current electricity.

There are two basic types of weld – **fillet welds** as shown in *figure 11.2a* and **butt welds** as shown in *figure 11.2b*. From a design point of view, butt welds are easily dealt with. A weld which passes through the whole thickness of the parent metal is a **full-penetration** butt weld, and provided the correct electrodes are used it can simply be assumed that the weld is at least as strong as the parent metal. The surface of the weld may subsequently be ground flush.

Fillet welds are specified in terms of **leg length**. This is the dimension shown in *figure 11.3*. Also shown is the **throat size**, and this is the dimension that determines the strength of the weld.

$$\text{Throat size} = \text{leg length} \times \cos 45°$$
$$\approx 0.7 \times \text{leg length}$$

Because of the importance of the throat size, a finished weld should always be convex in shape. The design strength of a fillet weld in Grade 43 steel is 215 N/mm^2. This figure already includes an allowance for the partial safety factor for material strength, γ_m. Therefore if we multiply the throat size by this stress we can produce a table of strengths for different leg lengths (*table 11.1*). Unless a weld forms a continuous loop, the length of a fillet weld should be reduced by two leg lengths to compensate for defective welding at the start and finish of the run.

Table 11.1
Strength of fillet welds

Leg length (mm)	Strength (kN/mm)
4	0.60
5	0.75
6	0.90
8	1.20
10	1.51
12	1.81
15	2.26

Example 11.1

Figure 11.4 shows an 80 × 80 × 8 angle welded to a gusset plate. It is to support a tensile design load of 300 kN. If it is welded all-round with a 6 mm fillet weld, what is the required length, *L*?

Solution

From *table 11.1* for a 6 mm fillet weld:

$$\text{Weld strength} = 0.9 \text{ kN/mm}$$

$$\text{Total length required} = \frac{300}{0.9} = 333 \text{ mm}$$

but

$$\text{total length} = 2 \times 80 + 2L$$

Answer $$\textbf{length, } \boldsymbol{L} = \frac{333 - 160}{2} = \textbf{86.5 mm}$$

11.4 Bolted connections

There are two basic types of bolt, and they each utilise a different basic principle to support a load:

- **Ordinary bolts** depend for their strength on contact between the bolt shank and the sides of the holes in the plates to be fixed – known as dowel-pin action. This is illustrated in *figure 11.5a*.
- **Friction grip bolts** are tensioned so that they clamp the plates together. Friction develops between adjacent faces (*figure 11.5b*). This produces a very rigid connection, and is therefore better for fixing members subjected to load reversal, such as wind bracing.

Bolts are generally fitted into holes which are 2 mm bigger than the bolt diameter. This allows a certain amount of adjustment when the steel is erected, and fabrication tolerances can thus be accommodated.

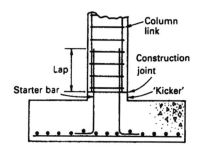

Figure 11.1k
The connection between a reinforced concrete column and foundation

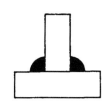

Figure 11.2a
A fillet weld

Figure 11.2b
A butt weld

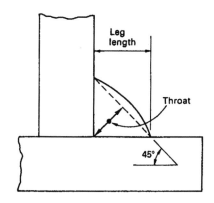

Figure 11.3
Key dimensions for fillet welds

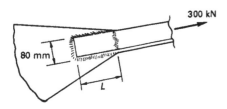

Figure 11.4
A fillet weld connection

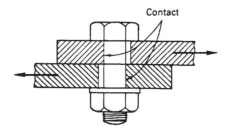

Figure 11.5a
Dowel-pin action of an ordinary bolt

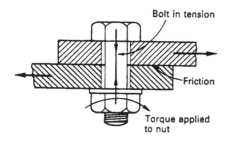

Figure 11.5b
Clamping action of a friction-grip bolt

11.4.1 *Design of ordinary bolts*

Ordinary bolts for structural use are available in two basic strengths or grades. Grade 4.6 are common mild steel, and sometimes referred to as **black bolts**. Grade 8.8 are higher strength. *Table 11.2* gives the design strengths for both grades. The values given include an allowance for the partial safety factor for material strength, γ_m.

Table 11.2
Strength of ordinary bolts (N/mm^2)

	Grade 4.6	Grade 8.8
Shear	160	375
Bearing	435	970*
Tension	195	450

*The bearing stress must be limited to 460 N/mm^2 when connecting Grade 43 steel.
Based on BS5950 Part 1

The design tensile capacity of a bolt is given by:

$$\text{Tension capacity} = \text{tensile strength} \times \text{tension area}$$

The tension area represents the area of the bolt at the root of the threads, and so is less than the nominal area of the bolt shank. For the commonly used sizes of standard metric bolts, the tension areas are given in *table 11.3*.

Table 11.3
Tension areas of metric bolts

Diameter (mm)	12	16	20	22	24	27	30
Tension area (mm^2)	84.3	157	245	303	353	469	561

Based on BS4190: 1967

Example 11.2

Figure 11.6 shows a column which is subject to a design uplift force of 200 kN. It is held down by four Grade 4.6 bolts cast into a concrete foundation. What bolt diameter is required?

Solution

$$\text{Load per bolt} = \frac{200}{4} = 50 \text{ kN}$$

From *table 11.2*, the tensile strength = 195 N/mm^2

$$\text{Tension area required} = \frac{50 \times 10^3}{195} = 256 \text{ mm}$$

From *table 11.3*

Answer   Use 22 mm diameter bolts (tension area = 303 mm^2)

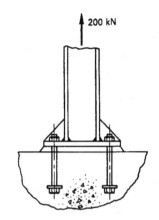

Figure 11.6
Column holding-down bolts

In most connections bolts are loaded in shear, and failure can take place in two possible ways. Firstly the bolt itself can shear as shown in *figure 11.7*. Secondly contact pressures between the bolt and the sides of the hole can produce a **bearing failure**. This either results in the hole in the plate becoming elongated or the bolt becoming indented. Both shear and bearing must be checked to determine the design load capacity.

$$\text{Shear capacity} = \text{shear strength} \times \text{bolt area}$$

In this case the bolt area can be taken as the full area of the bolt shank provided the threaded portion of the bolt does not cross the shear plane. Otherwise the tension area should be used.

$$\text{Bearing capacity} = \text{bearing strength} \times \text{bolt diameter} \\ \times \text{plate thickness}$$

The design load capacity of the bolt is the **minimum** of the two values obtained from the above formulae.

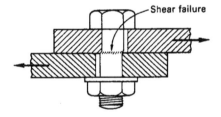

Figure 11.7
Failure due to shear in the bolt

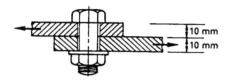

Figure 11.8a
A bolt in single shear

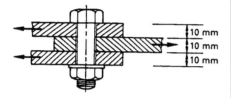

Figure 11.8b
A bolt in double shear

Example 11.3

Determine the design load capacity of the two bolted connections shown in *figure 11.8* if the grade 4.6 bolts are 20 mm diameter.

Solution

Case (a) – This is known as **single shear** as there is only one shear plane.

$$\text{Shear capacity} = \text{shear strength} \times \text{bolt area}$$
$$= 160 \times \pi \times 10^2 \times 10^{-3} = 50.3 \text{ kN}$$
$$\text{Bearing capacity} = \text{bearing strength} \times \text{bolt diameter} \times \text{plate thickness}$$
$$= 435 \times 20 \times 10 \times 10^{-3} = 87.0 \text{ kN}$$

Answer Load capacity = 50.3 kN

Case (b) – This is a case of **double shear**, because the bolt would have to shear across two planes before failure could occur. In double shear:

$$\text{Shear capacity} = 2 \times \text{shear strength} \times \text{bolt area}$$
$$= 2 \times 160 \times \pi \times 10^2 \times 10^{-3}$$
$$= 100.6 \text{ kN}$$

The bearing capacity of the inner plate is as *case (a)*, i.e. 87.0 kN.

Answer   Load capacity = 87.0 kN

11.4.2 *The design of friction grip bolts*

It has already been said that friction grip bolts work by clamping plates together to create friction. They are made from high strength steel. The tension in the bolt is achieved either by tightening the bolts to a specific torque, or by using **load indicating bolts** or washers. These contain small protrusions that yield as the bolt is tensioned (*figure 11.9*). The correct tension is determined by measuring the gap with a feeler gauge. The surface of the steel plates must be left unpainted in the region of the bolts so that friction is not affected.

Friction grip bolts can also fail in two ways. The joint can slip,

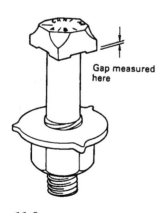

Figure 11.9
A load indicating friction grip bolt

or the bolt can fail in bearing, even though the bolt is not in contact with the side of the hole. A bearing failure, in this instance, refers to local yielding in the region of the bolt.

$$\text{Slip capacity} = 1.1\mu \times \text{bolt tension}$$

where μ is a slip factor (or friction coefficient), usually

$$\mu = 0.45$$

The bolt tension is usually assumed to be the proof load for the bolt, and values are given in *table 11.4*.

Table 11.4
Proof tension loads for friction grip bolts

Diameter (mm)	12	16	20	22	24	27	30
Proof load (kN)	49.4	92.1	144	177	207	234	286

Based on BS4395: Part 1

The bearing capacity is determined in the same way as for ordinary bolts, but the bearing stress for Grade 43 steel is increased to 825 2 N/mm^2.

Example 11.4

Determine the design load capacity of the bolted connection shown in *figure 11.8a* if the bolt is a 20 mm diameter friction grip bolt.

Solution

From *table 11.4* the bolt tension is 144 kN

$$\text{Slip capacity} = 1.1\mu \times \text{bolt tension}$$
$$= 1.1 \times 0.45 \times 144 = 71.3 \text{ kN}$$
$$\text{Bearing capacity} = \text{bearing strength} \times \text{bolt diameter}$$
$$\times \text{plate thickness}$$
$$= 825 \times 20 \times 10 \times 10^{-3} = 165 \text{ kN}$$

Answer   Load capacity = 71.3 kN

11.4.3 *Bolt spacing*

The load capacity of a group of bolts is simply the sum of the individual bolts. Where bolts are used in groups they must be separated by a certain minimum spacing. This is generally taken to be a minimum centre-to-centre spacing of 2.5 times the bolt diameter. Also bolts must not be too close to the edge of members. For standard rolled sections and plates the minimum distance from the centre of a hole to the edge of the member is 1.25 times the bolt diameter, and the minimum distance to the end of a member is 1.4 times the bolt diameter.

The above rules permit bolted gusset plate connections of framed structures to be designed in detail by simple scale drawing. It is assumed that an analysis has been carried out, so that the forces in all the members are known. The steps are as follows:

Step 1 – Draw lines representing the centroidal axes of the members meeting at a single point.

Step 2 – Superimpose the outlines of the actual members so that they are as close as possible to the intersection point. The position of the centroid for angle members is obtained from section tables – see *Appendix*.

Step 3 – Taking account of the member plate thicknesses, determine the load capacity of a single bolt, and hence the required number of bolts to resist the force in each member. If the result is not reasonable, change the type or diameter of bolt and repeat.

Step 4 – Starting from the end of each member, space the bolts according to the above rules.

Step 5 – Draw the gusset plate, which should be fabricated from material which is roughly the same thickness as the members being connected.

Example 11.5

In *example 8.4* we designed a simple crane structure, which consisted of two members meeting at B:

a 90 mm × 6 mm flat bar had a tensile design load of 103.5 kN;
a 90 × 90 × 6 angle had a compressive design load of 62.1 kN.

Detail a suitable gusset plate for the connection at B if 20 mm
diameter Grade 4.6 bolts are to be used with a 6 mm thick gusset
plate. Allow for a 25 mm diameter hole to connect the crane
winch.

Solution

$$\text{Shear capacity} = \text{shear strength} \times \text{bolt area}$$
$$= 160 \times \pi \times 10^2 \times 10^{-3} = 50.3 \text{ kN}$$

$$\text{Bearing capacity} = \text{bearing strength} \times \text{bolt diameter} \times$$
$$\text{plate thickness}$$
$$= 435 \times 20 \times 6 \times 10^{-3} = 52.2 \text{ kN}$$

$$\text{Bolt capacity} = 50.3 \text{ kN}$$

By inspection of the design loads in the members we can conclude
that the flat tie must have three bolts and the angle strut must have
two.

$$\text{bolt spacing} = 2.5 \times 20 - 50 \text{ mm}$$
$$\text{edge distance} = 1.25 \times 20 = 25 \text{ mm}$$
$$\text{end distance} = 1.4 \times 20 - 28 \text{ mm}$$

Answer   A possible solution is shown in *figure 11.10*

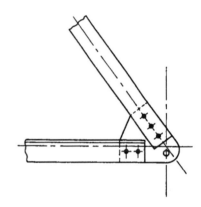

Figure 11.10
A gusset plate connection detail

11.5 Design of beam splices

Figure 11.1i showed a typical beam splice. It is assumed that the
flange plates, on the top and bottom of the beam, take all the
bending forces, and that the **web plates** support the shear. The
bending moment at the splice is therefore resisted by pure tensile
and compressive forces in the flange plates acting at a lever arm of
d. This is shown in the free-body diagram in *figure 11.11*. There
must be sufficient bolts to transmit the flange plate loads to the
beam flange. This, of course, puts a shear load on the bolts.
Similarly there must be sufficient bolts in the web plate to transmit
the shear force.

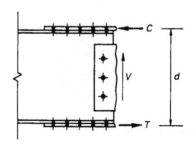

Figure 11.11
The free-body diagram of a beam splice

Example 11.6

A simply supported beam spans 15 m and is to be fixed inside an existing building. It is subject to the following loads:

$$\text{characteristic dead load} = 24 \text{ kN/m}$$
$$\text{characteristic imposed load} = 18 \text{ kN}$$

Because of access problems the beam must be split into three pieces of equal length.

a. Determine the size of a suitable standard universal beam in Grade 43 steel.
b. Design a suitable beam splice.
c. Design a suitable end-plate connection for fixing to existing steelwork.

Solution

$$\text{Design load} = (1.4 \times 24) + (1.6 \times 18)$$
$$= 62.4 \text{ kN/m}$$

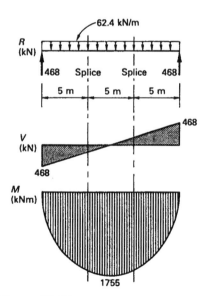

Figure 11.12a
Shear force and bending moment diagrams

a. $$\text{Reactions} = 62.4 \times 15/2 = 468 \text{ kN}$$

$$\text{Max. bending moment} = \frac{wL^2}{8} = \frac{62.4 \times 15^2}{8}$$

$$= 1755 \text{ kNm}$$

Figure 11.12a shows the resulting bending moment and shear force diagrams. For Grade 43 steel we can use a yield stress of 275 N/mm² (*table 3.4*).

$$\text{Reqd. plastic modulus, } S = \frac{M_{\text{ult}}}{\sigma_y} = \frac{1755 \times 10^6}{275 \times 10^3}$$

$$= 6382 \text{ cm}^3$$

Answer   Use 838 × 292 × 176 kg/m universal beam
(S = 6810 cm³)

b. From the free-body diagram shown in *figure 11.12b*:

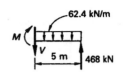

Figure 11.12b
The free body to one side of the splice

Shear force at splice, $V = 468 - (62.4 \times 5) = 156$ kN

Moment at splice $= (468 \times 5) - (62.4 \times 5 \times 2.5)$
$$= 1560 \text{ kNm}$$

The lever arm, d, is the distance between the centres of the flange plates. If we now estimate a plate thickness for the flange plates of 20 mm, and refer to section tables for the depth of the beam we get:

$$d = 835 + 20 = 855 \text{ mm}$$

$$\text{Flange force} = \frac{1560 \times 10^6}{855} = 1.825 \times 10^6 \text{ N}$$

$$\text{Flange area} = \frac{1.825 \times 10^6}{275} = 6635 \text{ mm}^2$$

Make the flange plate the same width as the beam, i.e. 292 mm.

It is proposed to use 24 mm diameter friction grip bolts as there is enough room for a double row on each side of the flange. The design single-shear load capacity for such bolts is 102 kN. *(Try proving this!)* The effective width of the flange plate must be found by deducting the effect of four 26 mm diameter holes

$$\text{Effective flange width} = 292 - (4 \times 26) = 188 \text{ mm}$$

$$\text{Plate thickness} = \frac{6635}{188} = 35.3 \text{ mm} \quad \text{say 35 mm}$$

$$\text{Number of flange bolts} = \frac{1825}{102} = 17.9 \quad \text{say 20 per side}$$

For the web plate use 14 mm thick plate (i.e. the same as the beam web).

$$\text{Number of web bolts} = \frac{156}{102} = 1.5 \quad \text{use say 4 per side}$$

Comment – There is a small additional shear force on the web bolts due to the moment caused by the horizontal spacing between the bolts. However, the use of the four bolts is safe.

Answer   A suitable beam splice is shown in *figure 11.12c*

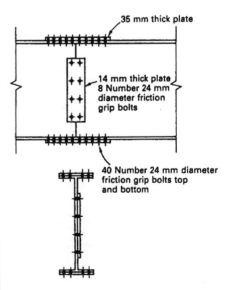

Figure 11.12c
The beam splice

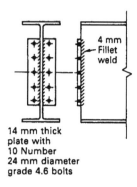

14 mm thick
plate with
10 Number
24 mm diameter
grade 4.6 bolts

Figure 11.12d
The end-plate connection

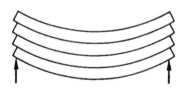

Cross-
section   Elevation

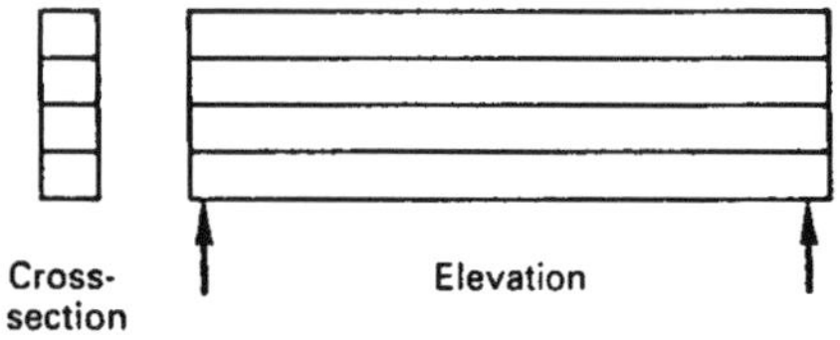

Figure 11.13a
An unloaded laminated timber beam

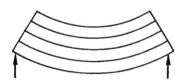

Figure 11.13b
Elevation of the beam after loading if the
laminates are unconnected

Figure 11.13c
Elevation of the beam after loading if the
laminates are securely bonded

c. Because the beam is simply supported, the bending moment is zero, and the end-plate connection must transmit only the shear force. This is, of course, equal to the reaction force = 468 kN.

Refering to *figure 11.12d*, and using 24 mm diameter Grade 4.6 bolts (design load capacity in single shear = 56.5 kN):

$$\text{Number of bolts} = \frac{468}{56.5} = 8.3 \quad \text{say } 10$$

Use an end-plate which is of similar thickness to the beam web, i.e. 14 mm. If the plate is 560 mm long:

$$\text{Load/mm on weld} = \frac{468}{2 \times 560} = 0.42 \text{ kN/mm}$$

From *table 11.1* it can be seen that a 4 mm fillet weld has a strength of 0.6 kN/mm, and would be suitable.

Answer   A suitable end-plate connection is shown in figure 11.12d

11.6 Connections in built-up beams

Any beam which is formed by connecting more than one component together is dependent, for its strength, on the quality of the fixing between the components. *Figure 11.13a* shows a laminated timber beam formed from four separate strips of wood. When a load is applied, if the strips are smooth and unconnected, the beam will deform as shown in *figure 11.13b*. In effect, it behaves like four independent shallow beams. On the other hand, if the laminates are securely glued together, it deforms as shown in *figure 11.13c*. It is now behaving like a unified deep beam, which, in this case, is four times stronger and eight times stiffer than the unconnected beam. The glue is necessary to transmit horizontal shear stresses from one laminate to the next.

The problem now is to develop a theory that allows us to evaluate this stress, and hence establish a required strength for the glue.

Derivation of formula for horizontal shear stress

Consider a short length of beam, δx, which is being subjected to a bending moment. At one end the moment is M, and at the other it is $M + \delta M$ (*figure 11.14a*). The beam can have any shape of cross-section. We know from the engineer's equation of bending that the moments will produce stresses in the beam:

$$\text{At one end} \quad \sigma = \frac{My}{I}$$

$$\text{At the other} \quad \sigma = \frac{(M + \delta M)y}{I}$$

Suppose we want the horizontal shear stress at any distance h from the neutral axis. Place an imaginary 'cut' at this level. We will consider the equilibrium of the free body above the 'cut' (*figure 11.14c*). It can be seen that, because of the unequal bending moments on the ends of the short length, there are bigger stresses pushing it one way than the other. In order to restore horizontal equilibrium we require a horizontal shear stress τ. We will therefore

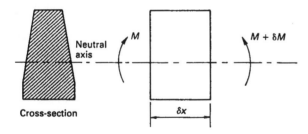

Figure 11.14a
A short length of beam with end moments

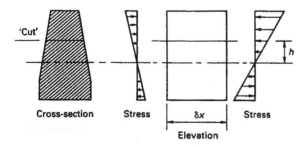

Figure 11.14b
Stresses produced by bending moments on the ends of the beam

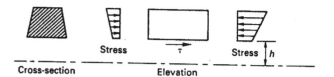

Figure 11.14c
Free body above the 'cut'

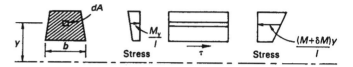

Figure 11.14d
τ must satisfy horizontal equilibrium

find the value of τ by evaluating the difference between the forces pushing on the left-hand and right-hand ends. The forces are obtained by integrating the stresses over the end areas above the 'cut'.

Consider a small elemental square of area dA at a distance of y from the neutral axis (*figure 11.14d*). The magnitude of the stresses at this level are given by the above two formulae. The force pushing on the left-hand end of the prism is therefore given by:

$$\text{Force} = \frac{My}{I} \times dA$$

Similarly the force pushing on the right-hand end is:

$$\text{Force} = \frac{(M + \delta M)y}{I} \times dA$$

If we take the difference between the above two expressions we get:

$$\text{Net force} = \frac{(\delta M)y}{I} \times dA$$

If we now integrate this over the whole area above h:

$$\text{Total net force} = \frac{\delta M}{I} \int y \, dA$$

The above integral is the first moment of area (of the region above h) about the neutral axis, say Q, and the expression can be rewritten:

$$\text{Total net force} = \frac{\delta M}{I} Q \qquad [11.1]$$

This force must be balanced by the force from the shear stress:

$$\text{Force from shear stress} = \tau b \, \delta x \qquad [11.2]$$

where

$$b = \text{width of beam at the 'cut'}$$

If we equate *[11.1]* and *[11.2]* and divide by b and δx we get:

$$\tau = \frac{\delta M}{\delta x} \frac{Q}{bI}$$

but as δx approaches zero we get the differential dM/dx which from *equation [7.1]* is the shear force V.

Hence

$$\textbf{shear stress, } \tau = \frac{\pmb{VQ}}{\pmb{Ib}} \qquad [11.3]$$

Also **shear flow**, q, is defied as shear force per unit length of beam:

$$\textbf{shear flow, } \pmb{q} = \frac{\pmb{VQ}}{\pmb{I}} \qquad [11.4]$$

Equation *[11.3]* is used to evaluate stresses in adhesives whereas equation *[11.4]* can be used to determine the size of welds in built-up beams. *Equation [11.4]* is also used to determine the spacing of individual connectors such as bolts, rivets, nails and spot welds. The following examples illustrate their use.

Example 11.7

Figure 11.15 shows a large built-up beam formed by welding together flat steel plates. It is to support a design shear force, V, of 3000 kN. Determine the required size of fillet weld.

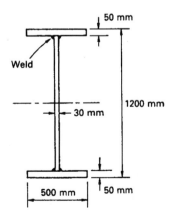

Figure 11.15
A steel built-up beam

Solution

The first step is to find the value of the second moment of area, I. From symmetry, the neutral axis is at the mid-height of the beam. Consequently the transfer inertial of the web will be zero. Also the self-inertia of the flanges will be negligibly small compared to the remaining two terms. The second moment of area therefore reduces to:

$$I_{NA} = I_{self(web)} + I_{transfer\,(flanges)}$$

$$= \frac{30 \times 1100^3}{12} + 2 \times 500 \times 50 \times 575^2$$

$$= 19.86 \times 10^9 \text{ mm}^4$$

From [11.4]   shear flow, $q = \dfrac{VQ}{I}$

$$= \frac{3000 \times 10^3 \times 500 \times 50 \times 575}{19.86 \times 10^9}$$

$$= 2171 \text{ N/mm}$$

but this must be divided between two welds:

$$\text{Shear flow/weld} = \frac{2171}{2 \times 10^3} = 1.085 \text{ kN/mm}$$

Answer
From *table 11.1*   **Use 8 mm fillet weld** *(capacity = 1.20 kN/mm)*

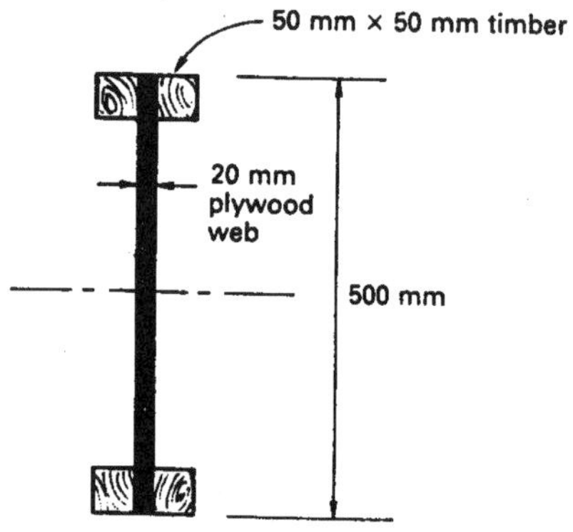

Example 11.8

The timber built-up beam shown in *figure 11.16* is subjected to a design shear force, V, of 5 kN.

a. If the flanges are attached by adhesive, what shear stress must it support?
b. If the flanges are attached by nails, and each nail can support a shear load of 100 N, what is the required nail spacing?

Figure 11.16
A timber built-up beam

Solution

As above
$$I_{NA} = I_{self(web)} + I_{transfer\ (flanges)}$$

$$= \frac{20 \times 500^3}{12} + 2 \times 2 \times 50 \times 50 \times 225^2$$

$$= 714.6 \times 10^6 \text{ mm}^4$$

a.

From [11.3]

Shear stress,
$$\tau = \frac{VQ}{Ib}$$

$$= \frac{5 \times 10^3 \times 50 \times 50 \times 225}{714.6 \times 10^6 \times 50}$$

Answer   Shear stress = 0.079 N/mm^2

b.
$$\text{Shear flow} = 0.079 \times 50 = 3.95 \text{ N/mm}^2$$

This means that, for every mm length of beam, 5.5 N of shear force must be transmitted. However, one nail will transmit 100 N.

Answer   Nail spacing $= \dfrac{100}{3.95}$ **= 25 mm**

11.7 Summary of key points from chapter 10

1. Connection failure is a common cause of structural collapse and so good detail design is very important. Unless good quality control is available on site it is best to reserve welding for the fabrication shop and make site connections with bolts.
2. **Rivets**, **bolts** and **welds** are the most common types of structural connectors for metals, whereas **screws**, **nails** and **glue** are common for timber.
3. Welds are commonly either **fillet welds** or **butt welds**. A full-penetration butt weld can be assumed to be as strong as the

parent metal. The strength of a fillet weld is related to the leg length.

4. **Ordinary bolts** depend upon dowel-pin action for their strength and it can be the shear strength of the bolt or the bearing capacity of the parent metal which is critical in determining the load capacity of a bolted connection. For each bolt in a connection:

$$\text{shear capacity} = \text{shear strength} \times \text{bolt area}$$
$$\text{bearing capacity} = \text{bearing strength} \times \text{bolt diameter} \times \text{plate thickness}$$

5. **Friction grip bolts** depend upon clamping plates together in order to produce friction between them. They can fail in bearing (but with a higher bearing strength than ordinary bolts) or by slipping:

$$\text{slip capacity} = 0.495 \times \text{bolt tension}$$

6. A beam can be joined or **spliced** with flange plates to carry the bending stresses and web plates to transmit the shear.

7. For **built-up beams**, which are fabricated from more than one component, it is necessary to check the shear capacity at the connection between the components so that the beam acts as a unified structural member.

8. Shear flow, $q = \dfrac{VQ}{I}$

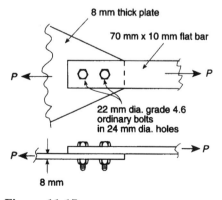

8 mm thick plate

70 mm x 10 mm flat bar

22 mm dia. grade 4.6 ordinary bolts in 24 mm dia. holes

8 mm

Figure 11.17

11.8 Exercises

E11.1 *Figure 11.17* shows a bolted connection between a flat mild steel bar ($\sigma_y = 275$ N/mm^2) and a steel plate. Determine the maximum permissible value of the design load, P, that the connection can support taking into account:

a) Tensile failure of the flat bar
b) Shear failure of the bolt
c) Bearing failure around the bolt
design load = 121.6 kN (shear of bolts)

E11.2 Repeat the previous example using 20 mm diameter friction grip bolts in 22 mm diameter holes.
design load = 132.0 kN (tension in flat)

E11.3 A 553 × 210 × 82 kg/m universal beam is strengthened by welding a steel plate 300 mm wide and 20 mm thick to its top flange as shown in *figure 11.18*. The composite section is subjected to a design shear force of 5000 kN. What size of continuous fillet weld is required at the join between the beam and the plate?
use 12 mm fillet weld

E11.4 *Figure 11.19a* shows the cross-section through a single track farm bridge which must be designed to support a single **design** axle load of 150 kN at any point. All other loads can be ignored. The bridge is simply supported and spans 8 m. The load can be assumed to be divided equally between the three beams.

a) Assume that the load is in the worst position for bending and determine the maximum tensile and compressive stresses in the compound section.
b) Assume that the load is in the worst position for shear and determine the shear flow on one weld in N/mm.

a) *stresses:* $\sigma_c = 25\ N/mm^2$, $\sigma_t = 90.3\ N/mm^2$
b) *shearflow* $q = 58.4\ N/mm$ *per weld*

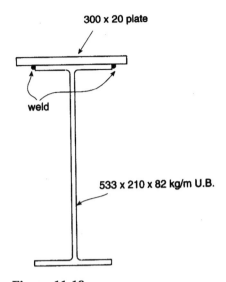

Figure 11.18
Composite beam made from universal beam and plate

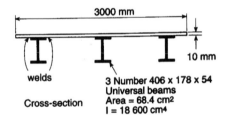

Figure 11.19a

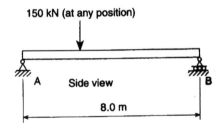

Figure 11.19b

Chapter 12

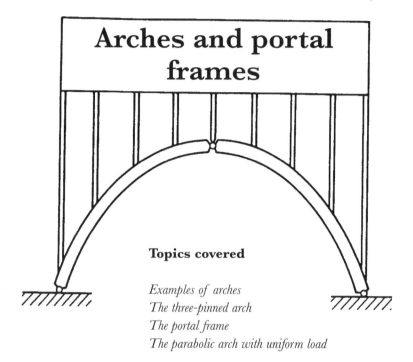

Arches and portal frames

Topics covered

Examples of arches
The three-pinned arch
The portal frame
The parabolic arch with uniform load

12.1 Introduction

The arch was first fully exploited structurally some two thousand years ago by the Romans, and many of their structures still survive. Arches featured extensively in the construction of the great medieval cathedrals, and were widely used throughout the industrial revolution for road and rail bridges. Arched structures can be particularly elegant and exciting, and modern materials have extended the scope for slender and stunning structures.

The reason why arches have been so extensively used throughout history is that they are a means of providing large clear open spans without the need for materials with a high tensile strength. Traditional masonry arches operate entirely in compression. A key feature of all arches is that their support reactions have a horizontal, as well as a vertical, component. The stability of an arch depends upon the ability of the supports to resist this horizontal component without excessive movement (*figure 12.1*). Arch foundation movement is the most frequent cause of collapse.

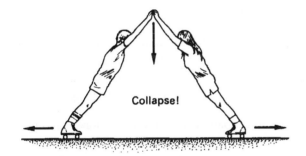

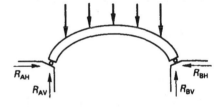

Figure 12.1
Arch supports must resist horizontal forces

A problem that we are faced with in this book is that most arches are too difficult to analyse. Throughout history, the design of masonry arches evolved empirically (and clearly only the successful ones have survived!). An accurate analysis of a typical Gothic cathedral is still beyond the capability of the most sophisticated computer techniques. The arches we use today tend to be much simpler in form, but are often statically indeterminate. To be determinate an arch must have three pins. This enables all the support reactions to be evaluated without taking into account the stiffness of the arch. We shall look at the analysis and design of three-pinned arches, and then go on to the special case of a parabolic arch with uniformly distributed load.

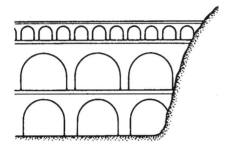

Figure 12.2a
A Roman aqueduct

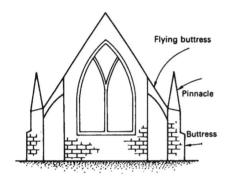

Figure 12.2b
The Gothic arch

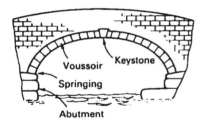

Figure 12.2c
An 18th-century canal bridge

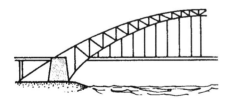

Figure 12.2d
An arch bridge with a suspended deck

12.2 Examples of arches

Figure 12.2a shows a typical Roman aqueduct structure. They were used to transport good quality water to their cities, and many are still surviving in good condition after nearly two thousand years. Timber **centring** or **falsework** was used to support the stones during the construction process.

Gothic arches are distinctive by the point at the apex. *Figure 12.2b* shows a much simplified form of this very ornate architecture. The great European cathedrals were built in this style during the 12th to the 15th centuries, but the style was copied extensively in the 19th century. The horizontal thrust from the arch was often transmitted by 'flying buttresses', which are particularly impressive in a no-tension material such as stone. Even the pinnacles have a function. They prestress the vertical buttresses so that they can resist more lateral force without tensile cracking.

A simple masonry canal bridge is shown in *figure 12.2c*. Many thousands of similar bridges survive traffic loads that were undreamed of when the bridges were designed. Some architectural terms are illustrated.

The Sidney Harbour Bridge shown in *figure 12.2d* is an arched bridge with a suspended deck. In many ways it is the inverse of a suspension bridge. Instead of a curved tensile cable it uses a curved compressive arch to support the deck load.

Figure 12.2e shows a simple three-pinned motorway footbridge. The three pins make the bridge statically determinate, and thus capable of easy analysis. Also a small amount of movement at the foundations can be tolerated as the stresses in the bridge are not affected.

A very common structural form is the portal frame (*figure 12.2f*). It is now almost universally used for large, cheap shed-type buildings for warehouses, factories and supermarkets. It can be designed with three pins, but the most economic form tends to have only two pins – one at each column foot. In practice the columns are usually bolted down to the foundation through a steel baseplate. You may think that this is not a pin, however, it is assumed to be so for design purposes. Only a small rotation of the foundation is required to lose any apparent fixity.

Finally *figure 12.2g* shows a dome. Domes can be thought of as three-dimensional arches. There is still a horizontal component to the reaction force around the perimeter of the dome, and this must be safely transmitted down to the ground. The method adopted by the Romans involved the use of heavy buttressing walls. However, in St Paul's Cathedral, Sir Christopher Wren ringed the dome with a tensile chain.

12.3 The three-pinned arch

Because a three-pinned arch is statically determinate it is relatively easy to analyse. Any point on an arch may be subject to a bending moment, a shear force and a direct thrust, and we can draw a diagram representing the distribution of each of these throughout the structure. The first step, as with beams, is to find the reactions. *Figure 12.3a* shows a three-pinned arch with a single point load, P.

(ΣM about A)	$P \times l = R_{BV} \times L$	hence R_{BV}
($\Sigma V = 0$)	$P = R_{BV} + R_{AV}$	hence R_{AV}

Because there is only a vertical load on the structure, we know that, for horizontal equilibrium, the horizontal reactions must be equal and opposite. To find their magnitude, consider a free body of any half of the arch, and take moments about the centre pin (*figure 12.3b*). We know that the bending moment must be zero at the pin.

(ΣM about C)	$R_{BH} \times H = R_{BV} \times L/2$	hence R_{BH}
($\Sigma H = 0$)	$R_{AH} = -R_{BH}$	

Also from *figure 12.3b* we can see by inspection that, for vertical and horizontal equilibrium, the shear at the centre pin must equal R_{BV}, and the thrust must equal R_{BH}.

To find the bending moment, shear force and thrust at any other point we consider a free body to one side of the particular point. Suppose we want the values at a distance x from A. From *figure 12.3c*, we use equilibrium to determine the unknown values of M, V and F, where V, the shear force, is the component of force perpendicular to the arch, and F, the thrust, is the component parallel to the arch. To find M:

(ΣM about 'cut')	$M = (R_{AV} \times x) - (R_{AH} \times y)$

The relative values of V and F change with variation in the slope of the arch. Knowing the angle of slope at the 'cut', we could use vertical and horizontal equilibrium to set up two simultaneous

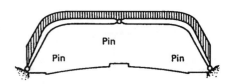

Figure 12.2e
A three-pinned arch footbridge over a motorway

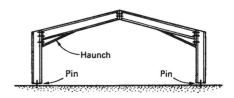

Figure 12.2f
A portal-framed factory building

Figure 12.2g
A dome is a three-dimensional arch

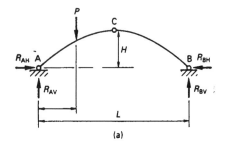

(a)

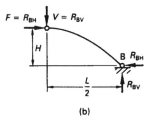

(b)

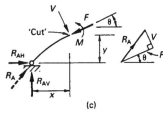

(c)

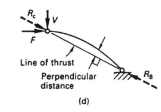

(d)

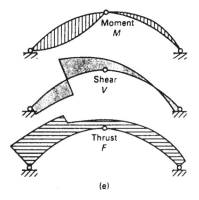

(e)

Figure 12.3
The three-pinned arch

equations, and solve them for V and F. Alternatively we could find the resultant, R_A, of R_{AV} and R_{AH}, and then draw a force triangle for V and F (*figure 12.3c*). If the distance, x, had taken us beyond the point where the load is applied, then the load must, of course, be taken into account when considering equilibrium. It is interesting to note that, for the unloaded half of the arch (*figure 12.3d*), the line of thrust of the resultant reaction force at B passes through the pin at C. The magnitude of the bending moment throughout BC is equal to the resultant force times the perpendicular distance to the arch.

Complete bending moment, shear force and thrust diagrams are shown in *figure 12.3e*.

Example 12.1

Draw bending moment, shear force and thrust diagrams for the arch shown in *figure 12.4a*. Indicate the magnitude of the principal values.

Solution

Reactions

($\sum M$ about A)	$25 \times 5 = R_{BV} \times 17$
	$R_{BV} = 7.35$ kN
($\sum V = 0$)	$R_{AV} = 25 - 7.35 = 17.65$ kN
($\sum M$ about D)	$7.35 \times 5 = R_{BH} \times 5$
	$R_{BH} = 7.35$ kN
($\sum H = 0$)	$R_{AH} = -R_{BH}$
	$R_{AH} = -7.35$ kN

The reactions are shown in *figure 12.4b*.

Moments

Because, in this case, the arch members are straight and the load is a point load, the diagrams will consist of straight lines.

$$M_C = (17.65 \times 5) - (7.35 \times 5)$$
$$= 51.5 \text{ kNm sagging}$$

Answer   The bending moment diagram is shown in
figure 12.4c

Shear forces

Consider the free body diagram shown in *figure 12.4d*:

Along AC

> shear force, $V = 17.65 \cos 45° - 7.35 \cos 45°$
> $= 7.28$ kN

$(\Sigma V = 0)$ – Along CD

> shear force, $V = 25 - 17.65 = 7.35$ kN

Answer   The shear force diagram is shown in *figure 12.4e*

Thrust

From *figure 12.4d* the sum of the parallel components is:

Along AC

> thrust, $F = 17.65 \sin 45° + 7.35 \sin 45°$
> $= 17.68$ kN

$(\Sigma H = 0)$ – Along CD

> thrust, $F = 7.3\,5$ kN

Along BD

> thrust $F = 7.35 \sin 45° + 7.35 \sin 45°$
> $= 10.35$ kN

Answer   The thrust diagram is shown in *figure 12.4f*

12.3.1 *The three-pinned arch with supports at different levels*

In the previous example, because the supports were at the same level, it was possible to find the vertical reaction at one support by taking moments about the other. The horizontal reactions could be ignored at this stage because they passed through the point which we were taking moments about. If the supports are not at the same level the horizontal reaction cannot be ignored. From *figure 12.5*:

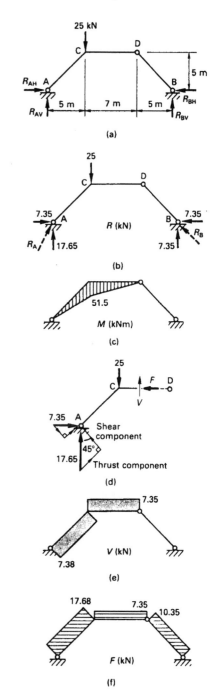

Figure 12.4

The three-pinned arch with a point load

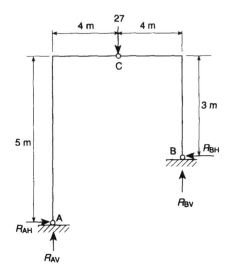

Figure 12.5
Arch with supports at different levels

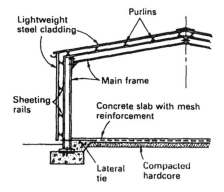

Figure 12.6
Typical portal frame details

(ΣM about A) $(27 \times 4.0) = (R_{BV} \times 8.0) + (R_{BH} \times 2.0)$
(ΣM about C) $(R_{BH} \times 3.0) = (R_{BV} \times 4.0)$

The above are simultaneous equations which yield:

$$R_{BH} = 13.5 \text{ kN}$$
$$R_{BV} = 10.13 \text{ kN}$$

Vertical and horizontal equilibrium can now be used to find the reactions at A.

Another approach, which avoids the need to solve a simultaneous equation, is to resolve all forces parallel to and perpendicular to a line which joins the supports A and B. However, in this case that is probably more work than the above.

12.4 The portal frame

The most economic form for a portal frame tends to have two pins, however, this is a statically indeterminate structure, and hence the analysis techniques are beyond the scope of this book. Nevertheless we now have enough knowledge to design a realistic and practical three-pinned portal-framed structure. The portal frame form is now almost universally adopted for large, open shed-type buildings.

Figure 12.6 shows typical construction details for a portal frame. Lightweight cladding is attached to the main frame by roof purlins and side sheeting rails. These are often manufactured from cold-rolled steel for maximum economy. For simplicity we shall assume that the loads are applied to the main frames as uniformly distributed loads, whereas they actually take the form of frequent point loads at each purlin and side rail location. This simplification makes very little difference to the final result.

As with all structures, the role of the designer is to anticipate all the possible failure mechanisms, and ensure that there is an adequate safety margin for all of them. As we saw in *section 4.5*, this usually involves checking several possible load cases using different partial safety factors for loads. For portal frames there are three load cases that are often critical. From *table 4.10*:

- Load case 1 – maximum vertical load = $1.4G_k + 1.6Q_k$
- Load case 2 – maximum sway = $1.2G_k + 1.2Q_k + 1.2W_k$
- Load case 3 – maximum uplift = $1.0G_k + 1.4W_k$

where G_k = characteristic dead load
           Q_k = characteristic imposed load
           W_k = characteristic wind load.

Example 12.2

A three-pinned portal framed building has a clear span of 15 m and the portals are spaced at 5 m intervals. The dimensions are shown in *figure 12.7a*. The building is subjected to the following loads:

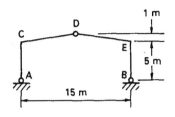

Figure 12.7a
Dimensions of the portal frame

Characteristic dead load of roof   = 0.25 kN/m²
Characteristic imposed load from snow = 0.6 kN/m²
Characteristic wind load   = 0.35 kN/m 2 on the
                                         windward wall
                                       = 0.15 kN/m² suction on
                                         the leeward wall

Ignore all other loads.
   Determine:

1. Bending moment and thrust diagrams for the vertical load and sway load cases.
2. A suitable standard section size for the main frame. (Assume that bracing will provide lateral restraint to the compression flange at 1.8 m spacing.)

Solution

Loading

Because the portals are spaced at 5 m intervals, each frame will carry a 5 m wide strip of roof.

Load case 1

$$\text{Design vertical load} = 1.4G_k + 1.6Q_k$$
$$= (1.4 \times 0.25) + (1.6 \times 0.6)$$
$$= 1.31 \text{ kN/m}^2$$
$$\text{Design UDL on portal roof} = 5 \times 1.31 = 6.55 \text{ kN/m}$$

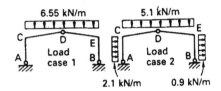

Figure 12.7b
Load cases

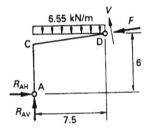

Figure 12.7c
The free body for case 1

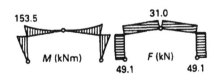

Figure 12.7d
Bending moment and thrust diagrams for
load case 1

Load case 2

$$\text{Design sway load} = 1.2G_k + 1.2Q_k + 1.2W_k$$
$$\text{Design load on roof} = (1.2 \times 0.25) + (1.2 \times 0.6)$$
$$= 1.02 \text{ kN/m}^2$$
$$\text{Design UDL on portal roof} = 5 \times 1.02 = 5.1 \text{ kN/m}$$
$$\text{Design UDL on windward wall} = 5 \times 1.2 \times 0.35$$
$$= 2.1 \text{ kN/m}$$
$$\text{Design UDL on leeward wall} = 5 \times 1.2 \times 0.15$$
$$= 0.9 \text{ kN/m}$$

See *figure 12.7b* for the two load cases.

Analysis

Load case 1 – the loading is symmetrical
Reactions – see *figure 12.7c*

$$(\Sigma V = 0) \qquad R_{AV} = R_{BV} = 6.55 \times 15/2 = 49.1 \text{ kN}$$

$$(\Sigma M \text{ about D})$$

$$R_{AH} = -R_{BH} = \frac{(49.1 \times 7.5) - (6.55 \times 7.5 \times 3.75)}{6}$$

$$= 30.7 \text{ kN}$$

Bending moments

$$M_C = R_{AH} \times 5 = 30.7 \times 5$$
$$= 153.5 \text{ kNm}$$

Thrust

$$\text{Slope of rafter} = \tan^{-1} 1/7.5 = 7.6°$$
$$(\Sigma H = 0) \qquad R_{AH} = F_{CD} \cos 7.6° = 30.7$$
$$F_{CD} = 31.0 \text{ kN}$$

Answer   see *figure 12.7d*

Load case 2

Reactions
Refer to f*igure 12.6a,b*

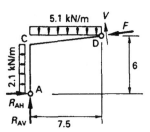

Figure 12.7e
Free body for load case 2

(ΣM about A)
$$R_{BV} \times 15 = (2.1 \times 5 \times 2.5) + (5.1 \times 15 \times 7.5) + (0.9 \times 5 \times 2.5)$$
$$R_{BV} = 40.8 \text{ kN}$$

($\Sigma V = 0$) $R_{AV} = (5.1 \times 15) - R_{BV} = 35.7 \text{ kN}$

From *figure 12.7e*

(ΣM about D)
$$R_{AH} \times 6 = (R_{AV} \times 7.5) - (5.1 \times 7.5 \times 3.75) - (2.1 \times 5 \times 3.5)$$
$$R_{AH} = 14.6 \text{ kN}$$

($\Sigma H = 0$) $R_{BH} = R_{AH} + (2.1 \times 5) + (0.9 \times 5)$
$$= 29.6 \text{ kN}$$

Bending moments

$$M_C = (R_{AH} \times 5) + (2.1 \times 5 \times 2.5)$$
$$= 99.3 \text{ kNm}$$

similarly $M_E = (R_{BH} \times 5) - (0.9 \times 5 \times 2.5)$
$$= 136.8 \text{ kNm}$$

Thrust

($\Sigma H = 0$) $R_{AH} + (2.1 \times 5) = F_{CD} \cos 7.6°$

$$F_{CD} = \frac{14.6 + 10.5}{0.9912} = 25.3 \text{ kN}$$

Answer   see *figure 12.7f*

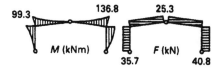

Figure 12.7f
Bending moment and thrust diagrams for load case 2

Design

It can be seen from the bending moment and thrust diagrams that load case 1 produces the highest values, and hence is the critical case. The worst combination, for which we must design the section, is:

$$\text{Design bending moment} = 153.5 \text{ kNm}$$
$$\text{Design thrust} = 49.1 \text{ kN}$$

Note – Although the vertical members are columns, the loading is dominated by the bending component. It is therefore more economical to use a universal beam section rather than a universal column.

We shall use *equation [11.2]* with the third term set to zero:

$$\frac{F}{Ap_{\text{c}}} + \frac{M_{\text{x}}}{p_{\text{b}}S_{\text{x}}} < 1.0$$

Try 406 × 140 × 39 kg/m universal beam
(A = 59.0 cm², S = 888 cm³, r_y 3.02 cm, x = 38.8)

$$\text{slenderness ratio} = \frac{L_{\text{E}}}{r_{\text{y}}} = \frac{1800}{30.2} = 59.6$$

from *figure 8.6*   compressive strength, $p_{\text{c}} = 215 \text{N/mm}^2$
from *figure 7.37*   bending strength, $p_{\text{b}} = 227 \text{N/mm}^2$

Therefore $Ap_{\text{c}} = 59.0 \times 215 \times 10^{-1} = 1268 \text{ kN}$
             $p_{\text{b}}S_{\text{x}} = 227 \times 888 \times 10^{-3} = 201.6 \text{ kNm}$

Substitute into *equation [11.2]*

$$\frac{49.1}{1268} + \frac{153.5}{201.6} = 0.80 < 1.0$$

Answer   Use 406 × 140 × 39 kg/m universal beam

12.5 The parabolic arch with uniform load

In *section 6.4.2* we saw that a suspended cable with a uniformly distributed load took up a parabolic shape. A cable, of course, cannot resist bending moments, and so must be in perfect tension. If we were to take the same parabolic shape and turn it upside down (*figure 12.8*), the resulting structure must be in perfect compression. Thus, for the special case of a parabolic arch with a uniformly distributed load, the bending moment throughout the arch is zero. Proof of this statement is given below. *Figure 12.9* shows two examples of bridges that can be approximated to the uniformly distributed case. However, individual heavy loads moving across the bridge will induce bending moments in the arch.

For design purposes we can use the suspended cable equations *[6.1]* to *[6.3]*:

$$\text{Vertical reactions, } R_{AV} = R_{CV} = \frac{wL}{2}$$

$$\text{Horiz. reactions, } R_{AV} = -R_{CV} = \frac{wL^2}{8D}$$

$$\text{Maximum force in arch} = \sqrt{(R_{AV}^2 + R_{AH}^2)}$$

To find the force in the arch at any intermediate point we can consider *figure 12.10*. We know from horizontal equilibrium that the horizontal component of force in the arch remains constant throughout. From vertical equilibrium we can see that at a distance *x* from the apex, the vertical component in the arch must be *wx*. The force in the arch is therefore simply the resultant of the two components:

$$\text{Force in arch at } x \text{ from apex} = \sqrt{((wx)^2 + (wL^2/8D)^2)} \quad [12.1]$$

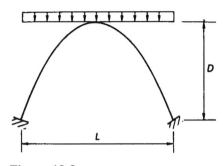

Figure 12.8
A parabolic arch with a uniformly distributed load

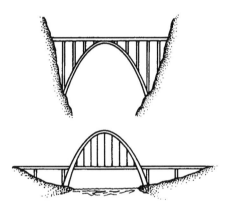

Figure 12.9
Two parabolic arch bridges with approximately uniform loads

Proof that a parabolic arch with uniform loading has no bending moments

The free body shown in *figure 12.10* is in vertical and horizontal equilibrium with the forces shown. If we can demonstrate that it is in moment equilibrium under those forces alone, then we know that there cannot be a moment at the cut.

We first need to find the distance *y*, and for this we need the equation of the particular parabola. The general equation of a parabola is:

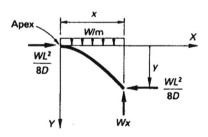

Figure 12.10
The free body of part of a parabolic arch

$$y = A + Bx^2$$

In our case at $x = 0$, $y = 0$

therefore $A = 0$

And at $x = L/2$, $y = D$

therefore $D = B \times \dfrac{L^2}{4}$

$$B = \dfrac{4D}{L^2}$$

So the equation is: $y = \dfrac{4Dx^2}{L^2}$ [12.2]

(ΣM about apex)

$$\text{clockwise moments} = \frac{wx^2}{2} + \frac{wl^2}{8D} \times \frac{4Dx^2}{L^2}$$

$$\text{cancelling} = \frac{wx^2}{2} + \frac{wx^2}{2}$$

$$= wx^2$$

$$\text{anticlockwise moments} = wx \times x = wx^2$$

Thus, by inspection moment equilibrium is satisfied without the need for a moment at the 'cut'.

12.6 Summary of key points from chapter 12

1. A true arch depends upon horizontal reaction forces at the supports for its stability.
2. A three-pinned arch is statically determinate and can therefore be analysed using the equations of equilibrium. Where the arch supports are at the same level the unknown reactions can be found as follows:
   2.1 find the vertical support reactions by taking moments about one support (as with a simply supported beam).

2.2 find the horizontal support reactions by taking moments about the central pin for one half of the arch.
3. When the supports are at different levels it is generally necessary to solve two simultaneous equations to obtain the unknown support reactions.
4. Once the reactions have been determined it is possible to produce shear, bending moment and thrust diagrams for the arch members.
5. A portal frame is a practical way of constructing large-span open buildings for factories, warehouses and supermarkets.
6. A parabolic arch with uniform loading is in perfect compression throughout (no bending). It is, in effect, the inverse of a parabolic tension cable.

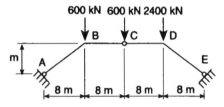

Figure 12.11
Three-pinned arch

12.7 Exercises

E12.1   *Figure 12.11* shows a three-pinned arch together with a set of design loads. Find:

a) The bending moments at B and D. State whether the top or bottom of the arch is in tension in each case.
b) The shear force and thrust on a cross-section in DE just to the right of D.
a) $M_B = 6000$ *kNm tension on top*, $M_D = 1200$ *kNm tension on bottom*
b) *thrust = 3590 kN, shear = 120 kN*

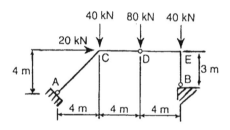

Figure 12.12
Three-pinned arch with horizontal load and supports at different levels

E12.2   For the three-pinned arch shown in *figure 12.12*

a) Find the vertical and horizontal support reactions at A and B.
b) Determine the bending moments at C and E stating whether the top or bottom of the arch is in tension.
c) Determine the value of the thrust and shear force along the member AC.
a) $R_{AV} = 54$ *kN*, $R_{AH} = 68$ *kN*, $R_{BV} = 106$ *kN*, $R_{BH} = 88$ *kN*
b) $M_C = 56$ *kNm*, $M_E = 264$ *kNm both tension in top*
c) *thrust = 86.3 kN, shearforce = 9.9 kN*

E12.3   The sway design load case for a portal frame is shown in *figure 12.13*. Determine the distribution of bending moment and thrust throughout the frame and hence obtain the maximum

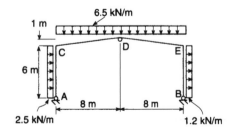

Figure 12.13
Sway loads on portal frame building

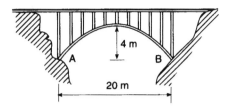

Figure 12.14
Parabolic arch bridge

bending moment and thrust values which must be used in the design of a suitable steel section.
moment = 210 kNm. thrust = 56.2 kN

E12.4   A bridge is supported by a parabolic arch as shown in *figure 12.14*. The arch itself is to be fabricated from two curved standard universal column sections suitably braced so that an effective length between bracing of 2 m can be assumed. The design load for each column section can be assumed to be 40 kN/m uniformly distributed across the entire span.

a) Determine the support reactions.
b) Calculate the maximum force in the arch.
c) Determine the dimensions of a suitable standard universal column section.

use 152 × 152 × 30 universal column

Chapter 13

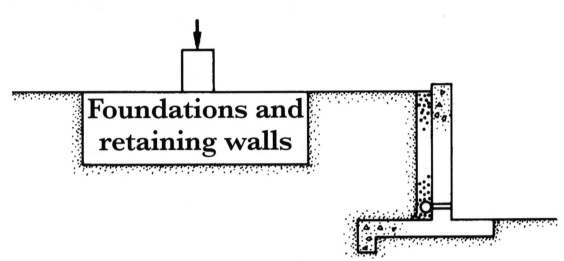

Foundations and retaining walls

Topics covered

Soil types and site investigation
Foundation types and selection
The design of unreinforced foundations
The design of pad foundations
Retaining wall types
The design of gravity retaining walls

13.1 Introduction

Foundations are an essential part of any building. Their purpose is to spread out concentrated structural loads from walls and columns onto the generally softer materials that form the surface of the earth's crust.

A **retaining wall** must be used when it is necessary to have a vertical change in ground level. The soil retained behind a wall produces a horizontal force that tries to push the wall over, and special calculations must be carried out to ensure that the wall is massive enough to prevent this happening.

Between them, foundations and retaining walls probably account for the largest number of structural failures. The failures result in cracked, leaning and bulging walls. They rarely cause fatal accidents, because they usually occur slowly, but they can be very expensive to rectify. The underlying cause of these failures is lack of adequate investigation of the site soil conditions and poor design and construction practices.

The design of both foundations and retaining walls involves considering the interface between the man-made structure and the naturally occurring soil. The study of soil strength and deformation, or **soil mechanics**, is a major subject in its own right, and we can only briefly consider it in this book.

13.2 Soil types

Soils are formed by the geological erosion of rocks, and consist of solid particles with spaces, or **voids**, between the particles. These voids can be filled with water or air, or a mixture of the two. However, in the UK most soils below a depth of about one metre are saturated (i.e. contain no air). It should be clear that the structural designer and the agriculturist have a very different definition of the word 'soil'. Top-soil, containing organic material, may be good for growing cabbages, but it is not chemically stable, and hence not regarded as a suitable material for use below a foundation. It must therefore be removed before construction begins.

Soils are categorised into six basic types according to the size of the particles:

- Boulders   greater than 200 mm
- Cobbles   200 mm to 60 mm
- Gravel   60 mm to 2 mm

- Sand   2 mm to 0.06 mm
- Silt   0.06 mm to 0.002 mm
- Clay   less than 0.002 mm

In practice, real soils usually consist of a mixture of the basic types and have descriptions such as 'sandy silt' or 'boulder clay'.

A soil which contains at least 25% clay particles behaves very differently from coarser soils. It sticks together, and is subject to swelling or contraction as its water content increases or decreases. This change in volume occurs whenever the stress on the soil changes. If such a soil is subject to an increase in stress, water is squeezed out of the soil. Consider the analogy of the wet sponge shown in *figure 13.1*. As pressure on the sponge is increased, more water is squeezed out and the volume of the sponge is reduced. The difference with clay soils is that this process can take a long time, as the voids between clay particles are very small. This process is known as **consolidation** and is a major cause of **settlement** of structures. A building can continue to settle for many decades as water is slowly expelled from the voids.

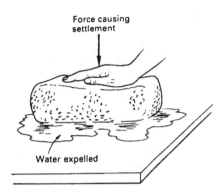

Figure 13.1
A wet sponge as an analogy to a clay soil under pressure

13.3 Site investigation

One of the first stages of any project is to commission a site investigation to determine the nature of the soil conditions. This will determine whether the site is suitable for the proposed structure, and enable the type and cost of the foundations to be estimated. Many economic and contractual problems arise because the site investigation is skimped and hence inadequate.

For small low-rise structures on virgin ground it may be adequate to carry out the investigation by means of **trial pits**. A small hydraulic excavator is used to dig a series of holes about 3 metres deep. These should be positioned just outside the perimeter of the proposed structure as shown in *figure 13.2*. The position of each hole and the nature of the soil encountered should be carefully recorded. The aim is to determine the suitability of the soil as a foundation material at various depths. The material should be carefully inspected as it is removed, and the case with which the excavator cuts through the soil is an important factor. For safety reasons, **no person must enter an unsupported excavation if it is more than one metre deep**. Usually the soil gets harder as the depth increases, and the ideal situation is to be able to identify a suitable firm bearing stratum within a metre or so of the finished ground surface.

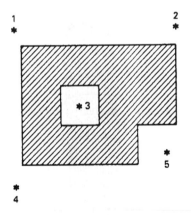

Figure 13.2
A building plan showing the proposed location of trial pits

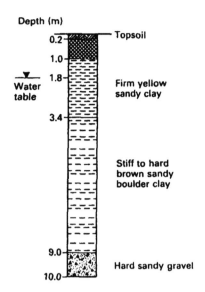

Depth (m)

0.2 — Topsoil
1.0 —
1.8 — Firm yellow sandy clay
Water table
3.4 —

Stiff to hard brown sandy boulder clay

9.0 —
10.0 — Hard sandy gravel

Figure 13.3
A typical borehold log

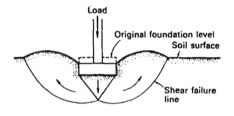

Figure 13.4
Bearing capacity failure of an axially loaded column foundation

Load
Original foundation level
Soil surface
Shear failure line

Figure 13.5
Differential settlement of the Campanile at Pisa

Where no suitable bearing stratum can be identified, and for large projects with heavy foundation loads, it is necessary to supplement the above trial pits with **boreholes** and field and laboratory tests. These are normally carried out by a specialist site investigation contractor, who will produce recommendations in the form of a site investigation report. A borehole consists of a hole drilled say 15 to 30 metres into the ground. Samples are extracted for examination and laboratory testing at regular intervals. The results are reported in the form of a borehole log (*figure 13.3*).

On wet sites, the level of the ground **water table** is of particular importance. This is the level at which water would form in an open hole. It can greatly increase the cost and difficulty of constructing deep foundations and basements. It can even cause basements and swimming pools to 'float' if they are not adequately held down.

Sites which occur in areas of mining, demolition, or waste tips require very special consideration, and it is often necessary to carry out research into the previous use of a site.

13.4 Foundations

A foundation must satisfy two design criteria. Firstly it must not cause the underlying soil to fail in shear – *figure 13.4* shows one particular failure mechanism. This is known as a bearing capacity failure. Secondly it must not be subject to excessive settlement – *figure 13.5* shows the world's most famous example. This is an example of differential settlement, where one part of a structure settles more than the other.

For small projects, and for the preliminary design of larger projects, the values of the permissible bearing pressures given in *table 13.1* can be used. These values should ensure that both the shear failure and the settlement criteria will be met.

Table 13.1
Permissible soil bearing pressures (kN/m^2)

Medium dense sand and gravel mix	300
Loose sand and gravel mix	150
Loose sand	75
Stiff clay	200
Firm clay	100
Soft clay	< 75

The above values are **permissible** pressures, not **ultimate** pressures; which means that they must be used in conjunction with **unfactored** loads.

We have already stated that the purpose of a foundation is to spread the load from hard structural walls and columns to relatively weaker soils. In *example 8.1* we designed the reinforcement for a concrete column which supported an unfactored load of 758 + 630 = 1388 kN. The column cross-section measured 325 mm square. This corresponds to a mean compressive stress of $1388/0.325^2 = 13\,140$ kN/m^2. If this value is compared to the permissible values given in *table 13.1*, it can be seen that a foundation must be used to spread the load over a much larger area.

The base of a foundation must not be too close to the finished ground level. This is to avoid the following problems:

- The soil near the surface is subject to seasonal variations in moisture content, and consequently its volume changes. Thus in summer the soil dries out and shrinks, and in winter it gets wet and swells. This is particularly a problem with clay soils. A depth of one metre to the underside of the foundation is normally adequate to avoid the seasonal variations.

- Some soils, particularly silts and fine sands, are subject to large increases in volume when frozen. This is known as **frost heave**. In most countries, even a sustained frost will not penetrate more than half a metre from the surface.

- Surface soil often contains organic material and the roots of vegetation which would disturb the foundation. (The roots of large shrubs and trees remain a problem, even at greater depths. Large trees should therefore not be planted close to buildings.)

13.4.1 *Examples of foundations and their selection*

Provided a suitable bearing soil can be found near the surface, the foundations for the walls of low-rise domestic houses are usually simple **strip foundations** or **strip footings** (*figure 13.6a*). They are normally unreinforced concrete, and should be constructed level, with vertical steps if necessary.

For columns supporting relatively low loads, simple unreinforced square or rectangular foundations can be used. These are referred to as **mass concrete pad foundations** (*figure 13.6b*).

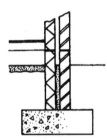

Figure 13.6a
A cavity wall on a strip foundation

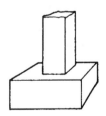

Figure 13.6b
An unreinforced pad foundation

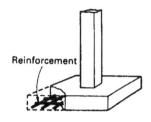

Figure 13.6c
A reinforced pad foundation

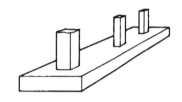

Figure 13.6d
A combined foundation

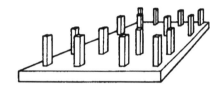

Figure 13.6e
A raft foundation

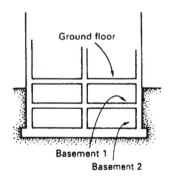

Figure 13.6f
A basement foundation

As loads increase, mass concrete foundations become less economic and **reinforced concrete pad foundations** are used (*figure 13.6c*). As we shall see below, the size of the pad depends upon the relationship between column load and the permissible ground bearing pressure.

If column loads increase to the point where the pad foundations are so big that the space between adjacent pads is less than the width of the pads themselves, then it becomes more economic to join the pads together to form a **combined foundation** (*figure 13.6d*).

Following the same logic, when the space between adjacent combined foundations becomes less than the foundation width, it becomes more economic to switch to a **raft foundation** (*figure 13.6e*). Raft foundations are also useful for lightly loaded structures over very weak soil.

If, even with a raft foundation, the bearing pressure is still too high, consideration must be given to distributing the load deeper within the soil mass. One way of doing this is by introducing one or more **basement** storeys (*figure 13.6f*). This works because the ground usually gets harder with increasing depth. The soil has already been subject to considerable load from the weight of over-burden soil. A good technique, used on some skyscrapers, is to make the weight of excavated soil the same as the weight of the new structure. The load on the soil at foundation level is thus unchanged.

Finally, the load can be distributed to deeper soil strata using **piles** (*figure 13.6g*). They can be either **bored piles**, which are cast *in situ* into pre-bored holes, or they can be **driven piles**, which are hammered into place. They resist load partly by means of friction on the pile surface, and partly by end-bearing on the tip of the pile. Piles are useful where soft soil overlies stronger material.

Example 13.1

Figure 13.7a shows a plan view of the column layout of a multi-storey building. Each column supports an unfactored axial load of 2000 kN. The soil at foundation level is stiff clay.
     Sketch a suitable form and size for the foundations.

Solution

From *table 13.1* for stiff clay

Permissible bearing pressure = 200 kN/m²

$$\text{Area of pad required} = \frac{2000}{200} = 10 \text{ m}^2$$

Thus a suitable pad would be

$$\sqrt{10} = 3.2 \text{ m}^2$$

Space between pads = 5.0 − 3.2 = 1.8 m

This is less than the width of the pads, so try a combined foundation for the whole row of five columns:

$$\text{Area required} = \frac{5 \times 2000}{200} = 50 \text{ m}^2$$

Thus a suitable combined foundation would be 22 m × 2.3 m (50.6 m²).

Answer   See *figure 13.7b*

13.4.2 *The design of unreinforced concrete foundations*

Lightly loaded foundations can be constructed without the need for reinforcement. The basic rule is that the depth of the foundation must be at least equal to the maximum projection of the foundation from the face of the column or wall. This is illustrated in *figure 13.8*.

Example 13.2

A cavity wall for a domestic building has an overall thickness of 275 mm. It supports a total unfactored load from the roof, floors and its own self-weight of 62 kN/m. The soil at foundation level is firm clay with a permissible bearing pressure of 100 kN/m².

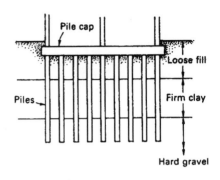

Figure 13.6g
Piled foundations

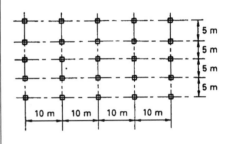

Figure 13.7a
The plan of the column layout

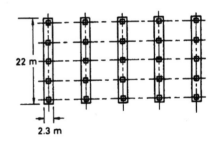

Figure 13.7b
The proposed foundation scheme

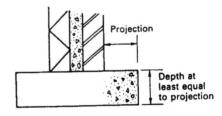

Figure 13.8
The proportions of an unreinforced foundation

Determine the dimensions of a suitable unreinforced strip foundation.

Solution

$$\text{Required width of strip} = \frac{62}{100} = 0.62 \text{ m}$$

$$\text{Projection from wall} = \frac{620 - 275}{2} = 173 \text{ mm}$$

Rounding up the above dimensions we get:

Answer   Use strip foundation 625 mm wide × 175 mm deep

13.4.3   *The design of reinforced concrete pad foundations*

A pad foundation can fail in several ways, and, as usual, it is the job of the designer to ensure that all possible failure mechanisms are checked. The size of the pad is determined from the unfactored column load and the permissible bearing pressure as shown in *example 13.1*. The remainder of the design process, which is concerned with reinforced concrete design, is carried out under limit state principles. We therefore use factored ultimate design loads. The total pad thickness will normally be about a sixth of its maximum plan dimension, but not less than 300 mm. We must determine the amount of tensile bending reinforcement and check for shear.

1. *Tensile reinforcement*

The foundation is designed as an inverted cantilever, with the ground pressure acting as an upwards load. The value of this pressure is the factored column load divided by the plan area of the pad. If the pad is of uniform thickness, its own self-weight can be ignored as it does not produce bending. The critical section for bending is assumed to occur at the face of the column. The required design moment must put the free body shown in *figure 13.9* into moment equilibrium.

Thus for a **square** pad with plan dimensions of length L supporting a **square** column of width C:

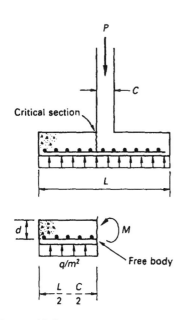

Figure 13.9

Bending in a pad foundation

Design column load = P

Ground pressure, $q = \dfrac{P}{L^2}$

Design moment $= q \times L \times \left(\dfrac{L}{2} - \dfrac{C}{2}\right) \times \dfrac{\left(\dfrac{L}{2} - \dfrac{C}{2}\right)}{2}$

$$= 0.5 \; qL \left(\dfrac{L}{2} - \dfrac{C}{2}\right)^2 \qquad\qquad [13.1]$$

An effective depth, d, is assumed and the tensile reinforcement is then determined as though the foundation was a rectangular beam of width L. Either *equation [7.17]* or, more economically, the design chart in *figure 7.46* can be used. For a square pad, identical reinforcement is used in the perpendicular direction. For small pads, the reinforcement can be uniformly distributed throughout the width of the foundation. However, if the width of the pad, L, exceeds $1.5(C + 3d)$, two-thirds of the reinforcement should be contained in a middle band of width $(C + 3d)$, as shown in *figure 13.10*. The rules about reinforcement spacing given in *section 7.9.3* apply throughout.

The reinforcement is, of course, placed in the tensile face of the foundation, which is the bottom. The foundation will also contain the starter bars for the column reinforcement as shown in *figure 11.1k*.

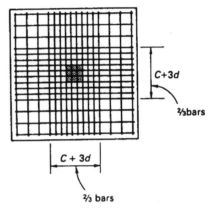

Figure 13.10
A plan view of the reinforcement layout

2. *Shear check*

The critical section for shear is assumed to occur $1.5d$ from the face of the column. Failure can occur either across the foundation or around the column as shown in *figure 13.11*. The latter case is known as **punching shear**. In either case it is only the ground force on the portion of foundation remote from the column that needs to be considered in shear calculations. This comes simply from consideration of vertical equilibrium of the free body.

Again for a square pad and column:

For shear across the foundation:

$$\text{Shear force, } V = q \times L \left(\dfrac{L}{2} - \dfrac{C}{2} - 1.5d\right)$$

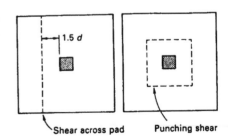

Figure 13.11
Critical sections for shear

$$\text{Shear area} = L \times d$$

$$\text{Shear stress, } v = \frac{V}{L \times d} \tag{13.2}$$

If v is less than the value of concrete shear stress v_c given in *table 7.1*, then no shear reinforcement is required. If v is greater than v_c, shear links may be provided but it is more common simply to increase d until the stress is reduced to less than v_c.

For punching shear:

$$\text{Shear force, } V = q \times (L^2 - (3d + C)^2)$$

$$\text{Shear area} = 4 \times (3d + C) \times d$$

$$\text{Shear stress, } v = \frac{V}{4 \times (3d + C) \times d} \tag{13.3}$$

Example 13.3

In *example 8.1* we designed a 325 mm square concrete column to support the following characteristic loads:

$$\text{Dead load} = 758 \text{ kN}$$
$$\text{Imposed load} = 630 \text{ kN}$$

Determine the dimensions and reinforcement for a suitable square pad foundation if it bears on stiff to firm clay with a permissible bearing pressure of 150 kN/m².

$$f_{cu} = 30 \text{ N/mm}^2, f_y = 460 \text{ N/mm}^2, \text{cover} = 50 \text{ mm.}$$

Solution

a. *Dimensions*

$$\text{Total unfactored loads} = 758 + 630 = 1388 \text{ kN}$$

$$\text{Required pad area} = \frac{1388}{150} = 9.25 \text{ m}^2$$

Width of pad = $\sqrt{9.25}$ = 3.04 m   say 3.1 m

Depth of pad ≈ 3.1/6 = 0.517   say 0.55 m

Answer   Use pad foundation 3.1 m square and 0.55 m deep

b. *Reinforcement*

Total design load, $P = 1.4G_k + 1.6Q_k$
$$= 1.4 \times 758 + 1.6 \times 630$$
$$= 2069 \text{ kN}$$

Ground pressure, $q = \dfrac{P}{L^2}$

$$= \frac{2069}{3.1^2} = 215 \text{ kN/m}^2$$

From *[13.1]*

Design moment = $0.5 \, qL(L/2 - C/2)^2$
$$= 0.5 \times 215 \times 3.1 \times (3.1/2 - 0.325/2)^2$$
$$= 642 \text{ kNm}$$

For concrete cover of 50 mm and assuming 16 mm diameter bars, the effective depth of the top layer of reinforcement is:

Effective depth, $d = 525 - 50 - 16 - 8 = 451$ mm

Using the design chart, *figure 7.46*

$$M/bd^2 = \frac{642 \times 10^6}{3100 \times 451^2} = 1.02$$

From chart $100A_s/bd = 0.29$
$$A_s = 0.29 \times 3100 \times 451/100$$
$$= 4055 \text{ mm}^2$$

Number of 16 mm bars = 4055/201 = 20.2

Check to see if bars can be uniformly distributed:

$$1.5(C + 3d) = 1.5(0.325 + 3 \times 0.451)$$
$$= 2.52 \text{ m} < 3.1$$

Therefore 2/3 of bars must be concentrated in a middle band.

$$\text{Width of band} = C + 3d = 0.325 + (3 \times 0.451)$$
$$= 1.68 \text{ m}$$

Number of bars in band = $2/3 \times 21 = 14$

Comment – 14 bars have 13 spaces between them.

Spacing within band = $1680/13 = 129$ mm

Place four bars each side of centre band at the maximum spacing of 160 mm.

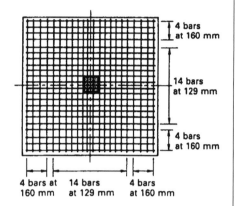

Figure 13.12
The layout of tensile bars for *example 13.3*

Answer   Use 22 number 16 mm diameter high yield bars in both directions as shown in *figure 13.12*

Shear

1. Shear across the pad:

$$\text{Shear force, } V = q \times L \times (L/2 - C/2 - 1.5d)$$
$$= 215 \times 3.1 \times (3.1/2 - 0.325/2 - 1.5 \times 0.451)$$
$$= 474 \text{ kN}$$

$$\text{Shear area} = L \times d = 3100 \times 451$$
$$= 1\ 398\ 100 \text{ mm}^2$$

$$\text{Shear stress, } v = \frac{V}{L \times d} = \frac{474 \times 10^3}{1\ 398\ 100}$$

$$= 0.339 \text{ N/mm}^2$$

From *table 7.1*

$$v_c = 0.49 > 0.339 \quad \text{satisfactory}$$

Punching shear:

Shear force, $V = q \times (L^2 - (3d + C)^2)$
$$= 215 \times (3.1^2 - (3 \times 0.451 + 0.325)^2)$$
$$= 1460 \text{ kN}$$

Shear area $= 4 \times (3d + C) \times d$
$$= 4 \times (3 \times 451 + 325) \times 451$$
$$= 3\,027\,100 \text{ mm}^2$$

Shear stress, $v = \dfrac{1460 \times 10^3}{3\,027\,100} = 0.482 \text{ N/mm}^2 < 0.49$

Proposed dimensions and reinforcement are satisfactory.

13.5 Retaining walls

We saw in *section 4.4* that soil produces a horizontal pressure on a wall, in a similar way to a liquid. The purpose of a retaining wall is to resist that force without excessive movement. The selection of the most appropriate type of retaining wall for a particular application is largely based on the height of soil to be retained. A range of retaining wall types, and the heights over which they are commonly used, is given in the next section.

13.5.1 *Examples of retaining walls*

Figure 13.13a shows a **mass** or **gravity** retaining wall. These can be constructed from mass concrete. brickwork or stonework. Historically they were used for large retaining walls adjacent to railway lines, but for modern structures the height is normally limited to three metres.

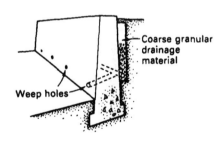

Figure 13.13a
A mass concrete gravity retaining wall

A **cantilever** wall is shown in *figure 13.13b*. These are mainly made from reinforced concrete, although reinforced brickwork or blockwork can be used. The weight of soil sitting on the base of the wall helps to provide stability. The 'heel' at the end of the base is to provide extra resistance to forward sliding. Cantilever walls are economic over the range from 2 metres to 8 metres.

As a means of reducing bending moments in high walls we can introduce **buttresses** in front of the wall or **counterforts**

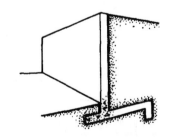

Figure 13.13b
A cantilever retaining wall (drainage and reinforcement not shown)

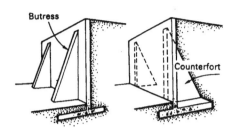

Figure 13.13c
Buttress and counterfort walls

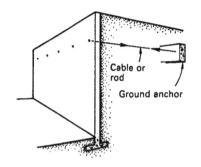

Figure 13.13d
An anchored retaining wall

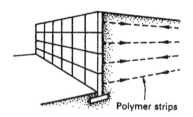

Figure 13.13e
A reinforced earth retaining wall

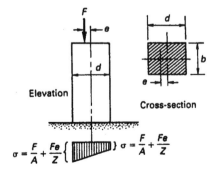

Figure 13.14
A lightweight block with eccentric load

behind the wall (*figure 13.13c*). These are placed at a spacing of say 1.5 times the height. They become economic for walls with a height over about 6 metres.

Another alternative for high walls is the **anchored** retaining wall shown in *figure 13.13d*. Here bending moments are reduced by a series of steel cables or rods which are anchored back to suitable bed-rock or **ground-anchors**.

A relatively recent development is the **reinforced earth** retaining wall (*figure 13.13e*). High strength polymer strip or matting is tied to the wall face at one end, and built into the compacted back-fill at the other.

13.5.2 *The middle-third rule for rectangular shapes*

Consider a block of material, with a rectangular cross-section, supporting a load which is eccentric to the centroidal axis (*figure 13.14*). Initially consider the material to be of negligible weight compared to the applied load, *F*. We know that both within, and underneath, the block, the stress distribution will be determined by *equations [9.3]* and *[9.4]*, the minimum stress being:

From *[9.4]*
$$\sigma_{min} = \frac{F}{A} - \frac{M}{Z}$$

where $M = Fe$. The question is – how great can the eccentricity, *e*, become before the minimum compressive stress is reduced to zero? The answer is – it can increase to the point where the bending component, Fe/Z, is equal to the axial component, F/A. If the block is *b* wide and *d* long, we know that for a rectangular cross-section:

and
$$A = b \times d$$
$$Z = b \times d^2/6$$

Therefore
$$\frac{F}{bd} = \frac{Fe}{bd^2/6}$$

hence
$$e = \frac{d}{6}$$

This means that provided the eccentricity, *e*, is less than $d/6$ from the centre of the base of the section, the stresses will remain compressive, i.e. **for no tension, the force must remain within the middle-third of the base of the member**.

If we now assume that the weight of the block, W, is significant compared to the applied force, F, we find that the stresses at the base of the block have an additional compressive component $= W/A$. In effect the mass of the block has introduced 'prestress'. This additional compressive stress means that the eccentricity of the applied force, F, can extend outside the middle-third before tensile stress results. This is shown in *figure 13.15*. We can replace the two vertical forces, W and F, with their resultant, R, at some eccentricity, e_r.

$(\Sigma V = 0)$ $\qquad\qquad R = W + F$

$(\Sigma M \text{ about } x)$ $\qquad\quad e_r \times R = F \times e$

hence $\qquad\qquad\qquad e_r = \dfrac{Fe}{(W + F)}$

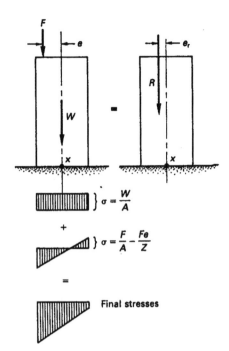

$$\sigma = \frac{W}{A}$$

$$\sigma = \frac{F}{A} - \frac{Fe}{Z}$$

Final stresses

Figure 13.15
A heavy block with eccentric load

It can be seen (*figure 13.15*) that this resembles the former case when the block was weightless, but the force has been replaced by the resultant. The rule now becomes – **for no tension, the resultant vertical force must remain within the middle-third of the base of the member**.

A further refinement is to replace the vertical applied force, F, with a horizontal one, acting at some height, h (*figure 13.16*). The bending component at the base of the block now becomes Fh. In order to satisfy horizontal equilibrium we also require an equal and opposite horizontal force at the base of the wall. In practice this requires adequate friction between the block and the foundation material. It is also possible to think of the moment from the force, F, as being exactly equivalent to a resultant force, R, applied at some eccentricity, e_r. In this case:

$(\Sigma V = 0)$ $\qquad\qquad\qquad R = W$

$(\Sigma M \text{ about } x)$ $\qquad\qquad e_r = \dfrac{Fh}{W}$

We can then apply the above middle-third rule to the resultant force.

Before we leave this section it is worth considering what happens if the resultant is allowed to stray outside the middle-third. Clearly

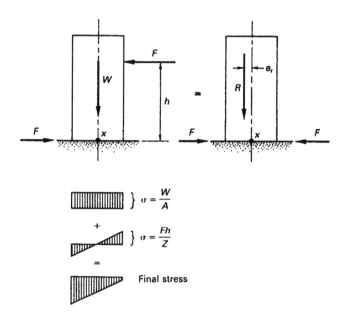

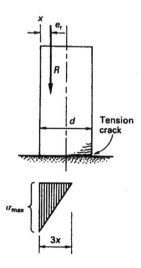

$$\sigma = \frac{W}{A}$$

$$+$$

$$\sigma = \frac{Fh}{Z}$$

$$=$$

Final stress

Figure 13.16
A heavy block with a horizontal load

when this occurs the tensile stress from bending exceeds the compressive stress from the self-weight. The application of *equation [9.4]* would indicate a region of tension at the base of the wall, which we have already indicated is not appropriate. It is assumed that a tension crack forms over the tensile region as shown in *figure 13.17*. The compressive stress forms a triangle. In order to satisfy both vertical and rotational equilibrium we know the following:

- The area of the stress triangle represents a force which must equal the vertical resultant force.

- The vertical resultant force must pass through the centroid of the triangle, which occurs at a point one-third of the way along the base of the triangle.

Therefore if $\qquad x = d/2 - e_r$ (*figure 13.17*)
Base length of triangle $= 3x$
Area of triangle $= 1/2 \times 3x \times \sigma_{max} = R$

Therefore $\qquad\qquad \sigma_{max} = \dfrac{2R}{3x}$

Figure 13.17
The resultant is outside the middle-third

The loss of tensile stress means that the above value of σ_{max} is greater than would have been obtained from *equation [9.4]*. We can therefore conclude that, when tensile stress cannot be supported, *equation [9.4]* can only be used if the vertical resultant force lies within the middle-third of the base.

However, the member has not yet collapsed, even though a tension crack has formed. A tension crack is unacceptable for two reasons. Firstly the material immediately above the crack is also put into tension, and hence may also crack. Secondly the repeated action of frost and debris entering the crack can cause progressive leaning of the structure.

As the position of the resultant nears the edge of the base, the value of σ_{max} increases significantly until it exceeds the compressive strength of either the structure or the foundation material (*figure 13.18*). As the material crushes it forms a pivot, and when the resultant vertical force moves beyond the pivot point, collapse results.

13.5.3 *Design of gravity retaining walls*

A gravity structure is one which depends for its stability on its own self-weight. Gravity retaining walls are often made from materials such as mass concrete and brickwork, which are weak in tension. For this reason they are usually required to be designed to have no (or very small) tensile stresses in the structure. Also they are often supported directly off the ground, and clearly the interface between the structure and the ground is not capable of transmitting long-term tensile stress. The usual requirement, therefore, is that the bearing pressures under gravity structures must remain compressive at all times.

There are probably more failures of retaining walls than any other type of structure. There are two principal reasons for this. Firstly, because of inadequate design, they are simply not built massively enough. Secondly, they are not well drained. If water can build up behind a wall, additional pressures from the water greatly increase the overturning forces on the wall. It is not surprising that many minor retaining walls collapse following a period of sustained heavy rain.

To evaluate the horizontal pressures from retained soil you are referred to *section 4.4.2*.

Gravity retaining walls are designed on traditional permissible stress principles rather than the more modern limit state

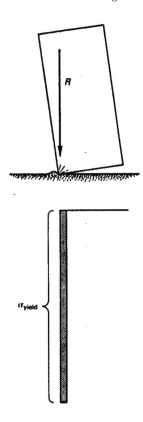

Figure 13.18
Collapse occurs when the resultant moves beyond the pivot point

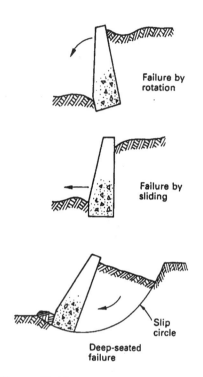

Figure 13.19

Failure mechanisms for a gravity retaining wall

approach. There are three common failure mechanisms – rotation about a point near the toe of the wall, sliding, and failure within the soil itself (known as a deep-seated failure (*figure 13.19*)). The aim of the design process is to ensure that there are adequate factors of safety against failure by any mechanism. The requirements are as follows:

- To prevent rotation about the toe, the vertical resultant force must be within the middle-third of the base, and the maximum bearing pressure must not exceed the permissible bearing pressure. Some typical permissible bearing pressures were given in *table 13.1*.

- To prevent sliding, friction is developed on the underside of the wall. In the absence of proper laboratory tests, a coefficient of friction of 0.6 can be assumed for *in situ* walls. Additional resistance to sliding is provided by **passive pressures** from the soil in front of the wall. Like active pressures these are also assumed to have a triangular distribution, and the maximum passive pressure can be taken as $k_p\gamma_s z$, where k_p is the **passive pressure coefficient**. In the absence of better information, this can be assumed to have a value of 3. The sum of the friction force and the passive resistance must exceed the horizontal force from the soil pressure. A factor of safety of 2 is required.

- Checking against a deep-seated failure is a soil mechanics problem, and hence beyond the scope of this book. It is mainly a problem on sloping hillsides, where retaining walls are often used to form terraces.

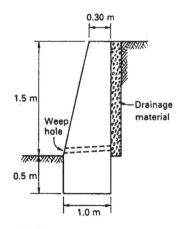

Figure 13.20a

A mass concrete retaining wall

Example 13.4

A well-drained mass concrete wall retains 1.5 m of soil as shown in *figure 13.20a*. The foundation material is firm clay. Using an active pressure coefficient, k_a, of 0.33, check that the dimensions of the wall are adequate. Assume

$$\gamma_{soil} = 20 \text{ kN/m}^3 \quad \text{and} \quad \gamma_{concrete} = 24 \text{ kN/m}^3$$

Solution

The passive pressures in front of the wall are ignored, as far as overturning is concerned, and the height of the wall is assumed to be 2.0 m. The first step is to locate the position of G, the centre of gravity of the wall. In all the subsequent calculations we shall consider a unit length of wall, i.e. 1 m.

Divide the cross-section of the wall up into three simple shapes, numbered 1–3 in *figure 13.20b*, and take moments about the point A for each shape:

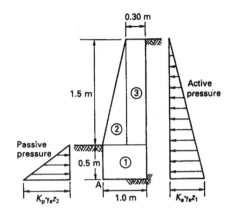

Shape	Weight (kN)		Distance (m)	Moment (kNm)
1	$0.5 \times 1.00 \times 24$	$= 12.0$	0.50	6.0
2	$1.5 \times 0.70 \times 24/2$	$= 12.6$	0.47	5.9
3	$1.5 \times 0.30 \times 24$	$= \underline{10.8}$	0.85	$\underline{9.2}$
		35.4		21.1

Figure 13.20b
Division of the wall into simple shapes, and the active and passive pressures on the wall

$$\text{Distance from A to G} = \frac{21.1}{35.4} = 0.596 \text{ m}$$

Distance from base centre $= 0.596 - 0.5 = 0.096$ m

Soil pressures – see figure 13.20b

From *section 4.4*

$$\text{Horizontal soil pressure at base} = k_a \gamma_s z_1 = 0.33 \times 20 \times 2$$
$$= 13.2 \text{ kN/m}^2$$
$$\text{Total horizontal force} = 13.2/2 \times 2 = 13.2 \text{ kN/m}$$
$$\text{Overturning moment} = 13.2 \times 2/3 = 8.8 \text{ kNm/m}$$

Check if resultant is within middle-third

We know that the magnitude of the vertical resultant must equal the total weight of the wall, W. To find the eccentricity of the vertical resultant force, take moments about the centre of the base – see *figure 13.20c*.

$$e_r = \frac{8.8 - (35.4 \times 0.096)}{35.4}$$
$$= 0.153 \text{ m}$$

Distance to edge of middle-third $= 1.0/6 = 0.166$ m > 0.153 m

Answer   Resultant lies within the middle-third

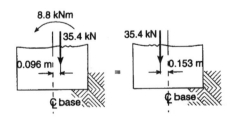

Figure 13.20c
Eccentricity of the resultant vertical force

Comment – You can see that even with such a massive wall, the resultant is only just inside the middle-third.

Evaluate maximum bearing pressure

From
$$\sigma_{max} = \frac{F}{A} + \frac{Fe}{Z}$$

$$= \frac{35.4}{1.0} + \frac{35.4 \times 0.153}{1^2/6}$$

$$= 35.4 + 32.5 = 67.9 \text{ kN/m}^2$$

In firm clay the permissible bearing pressure is 100 kN/m^2
$$> 73.7 \text{ kN/m}^2$$

Answer   Bearing pressure is acceptable

Check sliding

Horizontal force from soil $= 13.2 \text{ kN}$
Resistance from friction $= 35.4 \times 0.6 = 21.3 \text{ kN}$
Passive resistance pressure $= k_p \gamma_s z_2 = 3 \times 20 \times 0.5$
$\qquad\qquad\qquad\qquad\qquad = 30 \text{ kN/m}^2$
Passive force $= 30 \times 0.5/2 = 7.5 \text{ kN/m}$

$$\text{Factor-of-safety against sliding} = \frac{21.3 + 7.5}{13.3} = 2.16 > 2$$

Answer   Factor-of-safety against sliding is adequate. The dimensions of the wall are suitable.

13.6 Summary of key points from chapter 13

1. Foundation soils are usually softer than structural materials. A foundation is therefore required to spread the load from a structure so that it is transmitted safely into the earth's crust.
2. A site investigation must always be carried out to determine the nature of the soil below a structure. **Trial pits** and **boreholes** are excavated and samples sent to the laboratory for testing.

3. A foundation must not cause **shear failure** in the soil nor be subject to excessive **settlement**. The base of a foundation must not be too close to the ground surface.
4. Most foundations are concrete, and as the foundation load increases change from being simple unreinforced **pads** and **strips** to reinforced pads, strips or **rafts**. In problem soils, or for heavily loaded structures **piled** foundations may be required.
5. The depth of an unreinforced concrete foundation must be at least equal to the maximum projection of the foundation from the loaded column or wall.
6. Reinforced concrete foundations must be designed to resist bending failure and both simple shear and punching shear.
7. For gravity retaining walls the resultant vertical force must remain within the middle-third of the base of the wall to ensure that the whole of the base remains in compression.
8. In addition the wall must not slide forwards and the soil must not be subject to a deep-seated failure.

13.7 Exercises

E13.1   A 400 mm square concrete column supports the following characteristic loads:

$$\text{characteristic dead load} = 350 \text{ kN}$$
$$\text{characteristic imposed load} = 275 \text{ kN}$$

Determine the dimensions of a suitable unreinforced concrete pad foundation if the underlying soil is stiff clay.
say 1.8 m × 1.8 m × 0. 75 m deep

E13.2   Determine the dimensions of a suitable reinforced pad foundation as an alternative to the unreinforced pad in the previous example. Design a suitable arrangement of reinforcement given:

$$\text{concrete cover} = 40 \text{ mm}, f_y = 460, f_{cu} = 30 \text{ N/mm}^2$$

say 1.8 m × 1.8 m × 0.3 m with 12 mm bars evenly spaced in both directions in bottom face.

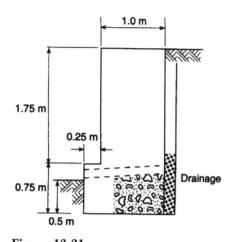

1.0 m

1.75 m

0.25 m

Drainage

0.75 m

0.5 m

Figure 13.21
Gravity retaining wall

E13.3   *Figure 13.21* shows a gravity retaining wall. Full drainage of water behind the wall can be assumed.

$$k_a = 0.33, \gamma_s = 20 \text{ kN/m}^3, \gamma_{concrete} = 24 \text{ kN/m}^3$$

Determine:

a) The maximum bearing pressure under the base of the wall.
b) The factor of safety against sliding.
σ_{max} = 88.7 kN/m², F.O.S. = 2.24

E13.4   How do the above answers change if the drainage behind the wall in the previous example becomes blocked and water rises to the top of the wall? *(Hint: refer to example 4.6)*
σ_{max} = 169 kN/m², F.O.S. = 1.32

Chapter 14

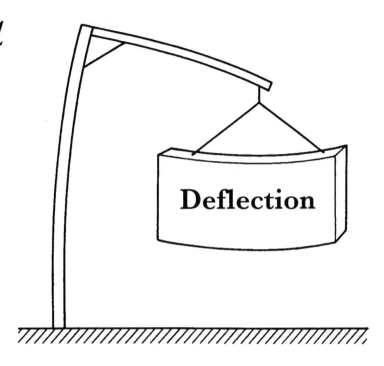

Topics covered

Permitted deflections
Deflections and limit state design
The deflection of pin-jointed frames
The deflection of beams

14.1 Introduction

Most of this book has been concerned with ensuring that structures do not fail through lack of strength. In limit state terms this is the ultimate limit state of collapse. However, one serviceability limit state which can influence the size of structural members is deflection. It is particularly a problem with long span beams, cantilevers and members made from material with low elastic modulus such as timber or fibre composites.

The assessment of deflection is essentially an analysis problem, and can be time consuming by hand. In all but the simplest of cases it is therefore modern practice to use a desk top computer for deflection calculations.

This chapter starts by considering what is an acceptable deflection for a structure. We then look at a method for calculating the deflection of pin-jointed frames. We go on to consider steel and concrete beams.

14.2 Maximum allowable deflection of structures

A structure which suffers excessive deflection is not acceptable because:

- Alarm and panic can be caused to users of a building if a structural member sags or sways too much in use.
- Damage can be caused to finishes such as plaster.
- Cracks can form which allow rainwater to penetrate and damage the structure.

Strictly speaking, deflections due to dead load can be compensated for, either by providing an upward camber to beams, or by introducing packing pieces (*figure 14.1*). (It is always good practice to give a small upward camber to prominent exposed beams for aesthetic reasons. A bridge beam which is actually level will unfortunately look as though it is sagging.)

Although different standards recommend slightly different values, in general the deflection, Δ, of beams should be limited to the span length divided by 360 if brittle finishes such as plaster are used. Otherwise the span divided by 200 is acceptable.

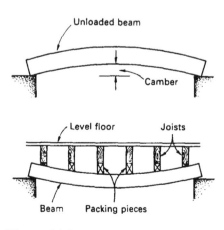

Figure 14.1
Methods of counteracting dead load deflection (shown exaggerated)

14.3 Deflection and limit state design

Excessive deflection is a serviceability limit state. This means that, when calculating deflections, we use the best estimate of the actual load on the structure and do not apply the safety factors that we use when checking the ultimate limit state of collapse. Therefore we use unfactored characteristic loads, which is of course the same as using a partial safety factor for loads, γ_f, of 1.0.

14.4 The deflection of pin-jointed trusses

Several methods are available for determining the deflection, Δ, of pin-jointed frames including graphical and computer-based techniques. We shall use a method suitable for hand calculation known as the **unit load method**. We shall first consider the steps involved. A proof of the method is then provided for those who are interested. We then go on to an example which illustrates the method.

   We shall assume that a pin-jointed frame has been analysed and suitable ties and struts designed. Clearly all those members which are in tension will get slightly longer, and all those in compression will get slightly shorter. We are interested in what effect these changes in length have on the deflection of a particular point in the structure.

Step 1 – Determine the force, F_L, in each member (tension or compression) that would result if the frame was subjected to unfactored characteristic loads.

Step 2 – Calculate the change in length of each member from

$$d = F_L L / EA$$

The extension of a tension member is considered positive whereas the shortening of a compression member is negative.

Step 3 – Apply a unit load to the point at which the deflection is required (i.e. a load of 1). The direction of the unit load should be in the direction of the desired deflection. Re-calculate the force, F_U, in each member of the structure due to the unit load. Tensile loads are positive and compression loads are negative.

Step 4 – For each member multiply the force F_U by its change in length under the real loads from *step 2*.

Step 5 – Sum up the result of *step 4* for all the members. The result is the desired deflection, i.e. $\Delta = \sum F_U F_L L / EA$.

The later steps are best carried out in tabular form.

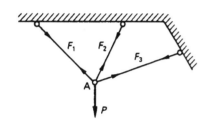

Figure 14.2a
A frame structure in equilibrium

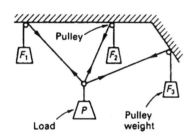

Figure 14.2b
Equivalent ropes and pulleys

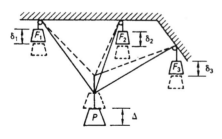

Figure 14.2c
A virtual displacement is applied

Proof of the unit load method

To prove the unit load method we shall use a very powerful theory known as the **principle of virtual work**. This is widely applicable to many analysis problems, but we shall restrict ourselves to the deflection of pinned frames at this stage. The principle is not easy to explain, but here goes!

We first need to define the concept of **work**. Work is said to have been done when a force moves through a distance. It therefore has units of Nm (1 Nm can be referred to as a **Joule**). Thus when an external load on a structure causes a deflection, it does some work. The application of the external load will also cause stresses and strains within the members of the structure. This is called **internal energy** (sometimes internal work), and if the principle of conservation of energy applies it is possible to equate external work with internal energy.

Consider the structure shown in *figure 14.2a*. (This particular structure is statically indeterminate, but the theory applies to any pin-jointed frame.) The joint at A is in equilibrium under the action of the load P and the forces in the three members – F_1, F_2 and F_3. Now imagine the structure has been replaced by the system of pulleys and ropes shown in *figure 14.2b*. If the load and the geometry are the same as before, then, for equilibrium, the pulley weights must exert the same forces in the ropes – F_1, F_2 and F_3.

Now suppose we move the load, P, by a distance Δ (*figure 14.2c*). This movement has not been caused by the load P or the forces in the ropes. It has been imposed by some external agent. For this reason it is known as a **virtual displacement**. Provided the displacement is small enough so that the angles of the ropes do not change significantly, the system will still be in equilibrium. The load has moved downwards but the pulley weights have moved upwards by δ_1, δ_2 and δ_3, respectively. By inspection of *figure 14.2c* we can see that the load, P, has done some work (i.e. force times distance). However, this must be exactly equalled by the negative work done by the pulley weights. We can therefore write:

$$P\Delta = F_1\delta_1 + F_2\delta_2 + F_3\delta_3$$

In a framed structure the displacements δ_1, δ_2 and δ_3 represent member extensions due to the virtual displacement (not the extensions due to the original loads). Also we could have many joints in a structure. The above equation can therefore be written:

$$P\Delta = \sum F\delta$$

The $\sum$ indicates that the product is summed for each member in the structure. This is the virtual work equation for a pin-jointed structure. Provided the force, P, and the displacement, Δ, act in the same direction, the term on the left-hand-side represents external work. The term on the right-hand-side represents internal energy stored (as in the weights that drive a grandfather clock). The important point to grasp is that the displacement, Δ, and the extensions, δ, are related to the geometry of the structure, but not directly to the load and forces. (Remember, we imposed the displacement on a structure which was already in equilibrium.) The equation therefore links two independent systems. One is a system of 'load and forces', the other is a system of 'deflection and extensions'.

Suppose, for the purposes of our solution, that the system of 'load and forces' consists of a single point load of unit magnitude, which is applied at the point where we want to know the deflection. Also it is applied in the direction of the desired deflection and it produces forces, F_U, in the members. The system of 'deflection and extensions' are the real unknown deflection, Δ, and the real member extensions, δ_L, under the real applied loads. The virtual work equation can therefore be written:

$$1 \times \Delta = \sum(F_U \times \delta_L) \qquad [14.1]$$

where Δ = the desired unknown deflection under the real loads
  F_U = member force under the unit load
  δ_L = member extension under the real loads.

From the consideration of strain in *chapter 2* we can easily show that the extension of a member, d, is given by:

$$\delta = FL/EA$$

Equation [14.1] can therefore be rewritten:

$$\Delta = \sum(F_U \times F_L L/EA) \qquad [14.2]$$

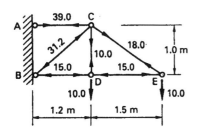

Figure 14.3a
Dimensions of the structure and member forces in kN

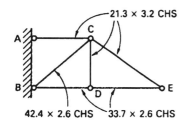

Figure 14.3b
Suitable standard circular hollow sections

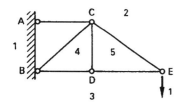

Figure 14.3c
The structure with unit load at E

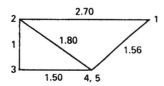

Figure 14.3d
The composite force diagram for the unit load

where F_L = member force under the real loads
L = length of member
E = modulus of elasticity
A = cross-sectional area of member.

This is the unit load method described above.

Example 14.1

In *examples 5.2* and *5.3* we analysed the forces in a pin-jointed cantilever bracket used to support pipes in a chemical plant. The forces we obtained are shown in *figure 14.3a*. *Figure 14.3b* shows some suitable standard steel circular hollow sections. Determine the deflection at the end joint E.

$$E_{steel} = 205\ 000\ \text{N/mm}^2$$

Comment – The loads given in example 5.3 were described as 'design' loads, which implies that partial safety factors for loads have been applied. However, for the purposes of this example we shall assume they are unfactored characteristic loads. Hence the forces given in figure 14.3a can be used directly in deflection calculations.

Solution

Step 1 – Has already been completed.

Step 2 – For each member we tabulate the length L, area A (from section tables), force F_L, and change in length δ:

Member	L (mm)	A (mm^2)	F_L (N)	$\delta = F_L L / EA$ (mm)
AC	1200	182	39 000	1.25
CE	1803	182	18 000	0.87
CD	1000	182	10 000	0.27
BD	1200	254	15 000	−0.35
DE	1500	254	15 000	−0.43
BC	1562	325	31 200	−0.73

Step 3 – A unit load is applied at E (*figure 14.3c*) and the structure re-analysed. *Figure 14.3d* shows the composite force diagram. The forces, F_U, are scaled off the diagram and the table extended:

Member	L (mm)	A (mm²)	F_L (N)	$\delta = F_L L / EA$ (mm)	F_U (mm)	$F_U\delta$
AC	1200	182	39 000	1.25	2.70	3.38
CE	1803	182	18 000	0.87	1.80	1.57
CD	1000	182	10 000	0.27	0	0
BD	1200	254	15 000	−0.35	−1.50	0.53
DE	1500	254	15 000	−0.43	−1.50	0.65
BC	1562	325	31 200	−0.73	−1.56	1.14
						$\Sigma = 7.27$

Answer   Deflection at point E = 7.27 mm

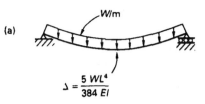
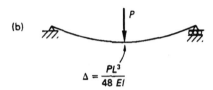
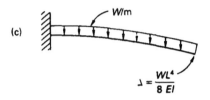
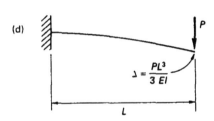
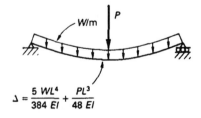

14.5 The deflection of beams

Formulae have been produced for calculating the deflection of uniform beams with simple patterns of loading. The majority of beams fit into one of the simple cases given in *figure 14.4*. The proof of one of these formulae is given below. Where a beam is subjected to a simple combination of loads, as shown in *figure 14.5*, we can use the principle of superposition and add the two cases. Also, for the purposes of design it is often good enough to approximate more complex loading patterns to one of the simple cases. *Figure 14.6* shows a loading pattern which can be approximated to case b in *figure 14.4*. *Figure 14.7* shows one which can be approximated to *case a*.

Figure 14.4
Maximum deflections for simple loading cases

Proof of the deflection formula for a point load at the mid-point of a simply supported beam

The unit load method, which we derived using virtual work, can be extended to include bending effects. We shall again use the virtual work equation, which states that external work is equal to internal work, to relate two independent systems. This time one system consists of 'load and bending moments' and the other consists of 'deflection and curvatures'. The first system comes from the unit load and the second system from the real loading. We saw from the engineer's equation of bending *[7.14]* that curvature, $1/R$, can be related to bending moment by:

Figure 14.5
Superposition applied to deflections

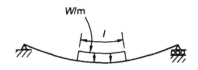

Figure 14.6
A load which can be approximated to a point load of wl

Figure 14.7
A load which can be approximated to a uniformly distributed load of $4P/L$

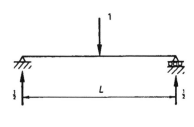

Figure 14.8a
The bending moment from a unit load

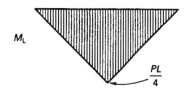

Figure 14.8b
The bending moment from applied load, P

$$\frac{1}{R} = \frac{M}{EI}$$

A problem is that, unlike pin-jointed members, which have constant forces and strains throughout, both the bending moments and curvatures usually vary along the length of a beam. We must therefore integrate over the whole beam. The unit load equation becomes:

$$\Delta = \int \frac{M_U M_L \, dx}{EI} \qquad [14.3]$$

where M_U = bending moment from the unit load
          M_L = bending moment from real applied load.

For our particular case, *figure 14.8a* shows the bending moment diagram for the unit load, and *figure 14.8b* shows the bending moment diagram for the applied load. We can see that, in this case, they are both triangular in shape. It is easy to show, by simple integration, that the area under the curve formed by multiplying the two triangles is $M_U M_L L/3$. This is known as a **product integral**, and values are tabulated in many text books for the commonly occurring bending moment diagram shapes such as triangles and parabolas. In this case, by substituting in *equation [14.2]*:

$$\Delta = \frac{L}{4} \times \frac{PL}{4} \times \frac{L}{3EI}$$

Hence $\Delta = \dfrac{PL^3}{48EI}$

Example 14.2

In *example 7.8* we determined that, based on strength, the maximum allowable span for a 50 mm wide × 150 mm deep timber joist was 3.67 m. The loading is shown in *figure 14.9*. Check the maximum deflection of the joist under full load, and compare it to the recommended allowable deflection if a plaster ceiling is to be used.

$$E_{\text{timber}} = 7000 \text{ N/mm}^2$$

Solution

Comment – Because timber design is still in this case based on permissible stress principles, the load is unfactored.

From *figure 14.4*

$$\Delta = \frac{5wL^4}{384EI}$$

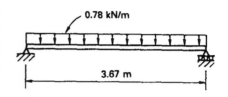

Figure 14.9
A timber floor joist

Great care must be taken with units. We shall work in N and mm even though it results in some very large numbers.

From *figure 7.36* the second moment of area, *I*, for a rectangular shape is:

$$I = bd^3/12 = 50 \times 150^3/12$$
$$= 14.062 \times 10^6 \text{ mm}^4$$

therefore

$$\Delta = \frac{5 \times 0.78 \times 3\,670^4}{384 \times 7\,000 \times 14.062 \times 10^6}$$

Answer   Deflection = 18.7 mm

Limit for a plaster ceiling = span/360 = 3670/360 = 10.2 mm

Conclusion – Even if we had considered only the imposed load acting, we can conclude that the deflection is excessive if the joist is used over its full span. Either the span should be reduced or a deeper joist used.

Example 14.3

In *examples 4.8* and *7.5* we designed two beams to support the roof of a double garage.

a. By referring to *example 4.9* show that the total unfactored loads produce the loading shown in *figure 14.10*.
b. Using the beam sizes determined in *example 7.5* (given below), calculate the mid-span deflection of each beam under this load.
   Beam B1 – 254 × 102 × 22 kg/m universal beam
   Beam B2 – 254 × 102 × 25 kg/m universal beam

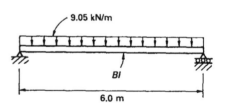

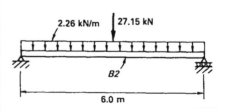

Figure 14.10
Unfactored loads on the double garage beams *B*1 and *B*2

Solution

a. *Loads*

Unit loads

$$\text{Total unfactored roof load} = 1.00 + 1.5 = 2.5 \text{ kN/m}^2$$
$$\text{Self-weight of beam} = 0.3 \text{ kN/m}$$
$$\text{Self-weight of handrail} = 0.07 \text{ kN/m}$$
$$\text{Dead load of walls} = 22.0 \times 0.215$$
$$= 4.73 \text{ kN/m}^2$$

Beam B1

Answer   UDL $= (2.5 \times 3.5) + 0.3 =$ **9.05 kN/m**

Beam B2

Answer   Design point load $= 9.05 \times 6/2 =$ **27.15 kN/m**

$$\text{UDL from brickwork} = 4.73 \times 0.4 = 1.89$$
$$\text{Handrail} = 0.07$$
$$\text{Self-weight} = \underline{0.3}$$

Answer   Unfactored UDL = 2.26 kN/m

b. *Deflections*

Beam B1
From section tables (see *appendix*) the second moment of area, *I*, for a $254 \times 102 \times 22$ universal beam is 2870 cm^4.

From *figure 14.4*
$$\Delta = \frac{5wL^4}{384EI}$$

$$= \frac{5 \times 9.05 \times 6000^4}{384 \times 200\,000 \times 2870 \times 10^4}$$

Answer   Deflection of *Beam 1* = 26.6 mm

Beam 2

From section tables (see *appendix*) the second moment of area, I, for a 254 × 102 × 25 universal beam is 3410 cm⁴

$$\text{For the UDL } \Delta = \frac{5wL^4}{384EI}$$

$$= \frac{5 \times 2.26 \times 6000^4}{384 \times 200\,000 \times 3410 \times 10^4}$$

$$= 5.6 \text{ mm}$$

$$\text{For the point load} \qquad \Delta = \frac{PL^3}{48EI}$$

$$= \frac{27.15 \times 10^3 \times 6000^3}{48 \times 200\,000 \times 3410 \times 10^4}$$

$$= 17.9 \text{ mm}$$

Answer   Deflection of *Beam 2* = 5.6 + 17.9 = **23.5 mm**

*Comment – Although both the deflection values exceed the recommended value of span/360, almost half the load is dead load, and hence packing, as shown in figure 14.1, can be used to counteract it. Nevertheless care must be taken to ensure that adequate **falls** are provided so that the roof does not collect rain-water.*

14.6 The deflection of reinforced concrete beams

The standard formulae from the previous section could be used to predict the deflection of reinforced concrete beams. A problem arises, however, when it comes to obtaining values for the elastic modulus, E, and the second moment of area, I. The value of E for steel is roughly seven times greater than that for concrete, and consequently the appropriate E value for a reinforced concrete beam will vary with the amount of reinforcement. Also the concrete below the neutral axis is assumed to be at least partially cracked. The question therefore arises as to how much concrete is included when the I value is calculated.

In most cases these awkward questions are avoided by simply ensuring that the ratio of a beam's span to its effective depth does not exceed a recommended value. For rectangular beams and slabs, the basic span/effective depth ratios are as given in *table 14.1*.

Table 14.1
Basic span/effective depth ratios

Cantilever	7
Simply supported	20
Continuous	26

Based on BS 8110: Part 1: 1997

The basic figures in *table 14.1* should then be modified to take account of the amount of tensile reinforcement and its stress. Where reinforcement is economically designed, so that it is close to its characteristic strength at its design ultimate moment, the factors shown in *table 14.2* should be used:

Table 14.2
Modification factor for tension reinforcement

f_y	M_{ult}/bd^2				
(N/mm^2)	0.50	1.00	2.00	4.00	6.00
250	1.90	1.55	1.20	0.94	0.82
460	1.56	1.30	1.04	0.84	0.76

Based on BS 8110: Part 1: 1997

Example 14.4

In *example 7.13* we designed a reinforced concrete beam to support the roof of a double garage. Given the following data, determine whether the deflection of the beam will be acceptable:

$$\text{Beam span} = 6 \text{ m (simply supported)}$$
$$\text{Beam dimensions} = 160 \text{ mm wide} \times 287 \text{ mm effective depth}$$
$$f_y = 460 \text{ N/mm}^2$$
$$M_{ult} = 61.74 \text{ kNm}$$

Solution

Table 14.1   Basic span/depth ratio = 20

$$M_{ult}/bd^2 = \frac{61.74 \times 10^6}{160 \times 287^2} = 4.7$$

Table 14.2   Factor = 0.84 (by interpolation)

Required span/depth ratio = 20 × 0.84 = 16.8

Actual span/depth ratio = 6 000/287 = 20.9

Answer   The effective depth should be increased to say 360 mm

14.7 Summary of key points from chapter 13

1. An important serviceability limit state is deflection. In general a beam should not deflect more that span/360 if brittle finishing materials are used, or otherwise span/200 is acceptable.
2. A partial safety factor for loads, $\gamma_f = 1.0$ is used for deflection calculations.
3. The deflection of a pin-jointed truss can be determined by the **unit load method** which is derived using the **principle of virtual work**.
4. The deflection of beams can also be determined using the unit load method, but in practice standard formulae can often be adequate:

	point load	**UDL**
simply supported beam	$\dfrac{PL^3}{48EI}$	$\dfrac{5wL^4}{384EI}$
cantilever beam	$\dfrac{PL^3}{3EI}$	$\dfrac{wL^4}{8EI}$

5. Deflections can be added using the **principle of superposition** provided that the structure remains within its elastic range.

6. For concrete beams, deflections are normally controlled by limiting the span/effective depth ratios to recommended values.

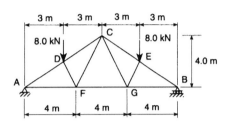

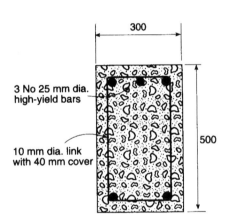

Figure 14.11
Timber roof truss

Figure 14.12
Cross-section of reinforced concrete cantilever beam

14.8 Exercises

E14.1   *Figure 14.11* shows a timber roof truss subject to two unfactored loads of 8 kN. All the members are 100×50 softwood with a modulus of elasticity, $E = 7000$ N/mm^2. Calculate the deflection of the point G.
deflection = 8.04 mm

E14.2   A $406 \times 178 \times 54$ kg/m steel universal beam is used as a simply supported beam span with a span of 7.5 m. It supports unfactored loads consisting of a uniformly distributed load of 16 kN/m and a point load at mid-span of 32 kN. Calculate the mid-span deflection.
deflection = 24.7 mm

E14.3   An aluminium flagpole is 6 m high and is made from circular tube with an outside diameter of 120 mm and a wall thickness of 2.0 mm. In high wind a flag can be considered to produce a horizontal point load at the top of the pole of 150 N. How much does the top of the pole deflect?
deflection = 120 mm

E14.4   *Figure 14.12* shows the cross-section of a reinforced concrete **cantilever** beam which is designed to resist a hogging bending moment of 190 kNm. What is the maximum length of the cantilever to ensure that its deflection will not be excessive?
maximum length = 2.9 m

Chapter 15

Indeterminate structures and computers

Topics covered

Examples of indeterminate structures
Standard cases
Upper and lower bound solutions
The 'stiffness' computer method
Introduction to the finite element method

15.1 Introduction

The earlier chapters of this book have largely been concerned with what are known as **statically determinate** structures. These are structures that contain just sufficient members and supports to carry the loads imposed. If a member or support is removed, such a structure will turn into a mechanism and collapse. On the other hand if an additional member or support is added the structure becomes **statically indeterminate** with each additional member or support reaction being a **redundancy**. A statically indeterminate structure is often referred to as a redundant structure.

Section 2.5 explains that a plane structure generally requires its supports to provide a minimum of three reaction forces to remain stable (say a pin support plus a roller support). *Section 7.1* indicates that both the simply supported beam and the cantilever beam contain this minimum number. The addition of extra supports to a beam will cause that beam to become statically indeterminate. *Section 5.4* indicates how to identify redundant members in pin-jointed trusses. *Section 12.1* explains that arches and portal frames are statically determinate when they contain three pins.

We have already seen that the analysis of statically determinate structures can generally be performed relatively simply by using the equations of equilibrium ($\Sigma H = 0$, $\Sigma V = 0$ and $\Sigma M = 0$). These equations enable a structure's reaction forces to be evaluated as well as the distribution of forces and bending moments throughout each member. The analysis of statically indeterminate structures is usually more difficult. It often involves knowing the **stiffness** of members within a structure. In the past, students of 'the theory of structures' have spent significant time studying a wide range of hand techniques and methods for the analysis of different types of statically indeterminate structure. Such manual techniques tend to get more onerous as the number of redundancies increases, and they often involve the need to make simplifying assumptions. Many of these techniques have now been replaced by computer-based methods, and the power of modern computers means that it is no longer necessary to make simplifying assumptions.

15.2 Examples of statically indeterminate structures

We have already seen in *Chapter 5* that crossed members often indicate redundancies in pin-jointed frames. The bridge truss *(figure*

15.1a) contains three redundancies, two from the crossed members and one from the extra roller support.

A simple form of statically indeterminate beam is the **propped cantilever** (*figure 15.1b*). The simple support at the end provides one redundancy. The extra supports in the **continuous beam** (*figure 15.1c*) add two redundancies. Any slight settlement of the supports would change the distribution of bending moments and shear forces throughout these beams.

The **two-pinned portal** (*figure 15.1d*) is one of the most commonly used structures for modern factories, warehouses and supermarkets. Such structures tend to contain less steel and hence be more economical than the three-pinned determinate version. Fully fixed portals (no pins) are less popular because of the difficulty of adequately fixing the bases of the portal columns. Only a very small rotation of the column foundations would be required to lose the effect of fixity.

The **Vierendeel girder** bridge (*figure 15.1e*) is a highly redundant structure that gains its strength from the stiffness of its members and the rigidity of its joints (as an alternative to the use of diagonal members in a pin-jointed frame).

Real structures for large buildings are rarely built with true pinned joints, and it is often possible to make economic savings in the use of materials if some allowance is made for the rigidity of the joints. Complex structures (*figure 15.1f*) can be designed by making simplifying assumptions or, more precisely, by using computer methods.

Curved **shell structures** provide a particular challenge to the designer. *Figure 15.1g* shows a shell supported by tension cables that forms part of the roof of a football stadium. A special computer technique known as the **finite element method** is appropriate in such cases. This technique can be used to analyse structures of any shape and mechanical engineers use it for stressing car bodies for example.

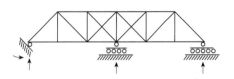

Figure 15.1a
A bridge truss with three redundancies

Figure 15.1b
A propped cantilever

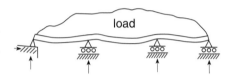

Figure 15.1c
A continuous beam with two intermediate supports

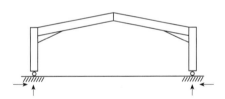

Figure 15.1d
A two-pinned portal

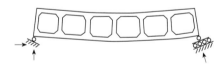

Figure 15.1e
A Vierendeel girder with rigid joints

15.3 Standard cases

For commonly occurring indeterminate structures many text books contain standard cases that enables the distribution of

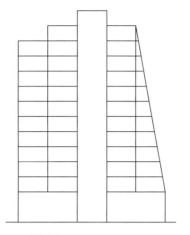

Figure 15.1f

A complex building frame that may be subjected to computer analysis

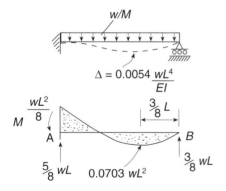

Figure 15.1g

A curved shell structure supported by tension cables

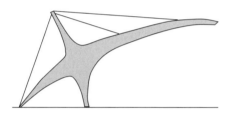

Figure 15.2a

Propped cantilever uniformly distributed load

bending moments and shear forces to be determined. Some of these for simple indeterminate beams are shown in *figure 15.2*. *Figure 15.2a* shows a propped cantilever with a uniformly distributed load (UDL) throughout. *Figure 15.2b* shows the same structure with a central point load. *Figure 15.2c* shows a beam fully fixed at both ends with a UDL and *figure 15.2d* shows the same beam with a central point load. In each case the top diagram shows the end conditions, deflected form and maximum deflection. The second diagram shows the peak bending moments, the bending moment diagram and the support reactions. Note that the general rule that UDLs produce curved bending moment diagrams, and that point loads produce straight lines, still applies to indeterminate beams.

15.4 Indeterminate structures and equilibrium

The propped cantilever shown in *figure 15.2a* only contains one redundancy and hence if only one of the unknown moments or reactions is known, the others can be determined by considering simple equilibrium. Supposing, for example, we knew that the support moment was $wL^2/8$, it is easy to derive the end support reaction, R_B:

(ΣM about fixed support)

$$\text{anticlockwise moments} = \text{clockwise moments}$$

$$\frac{wL^2}{8} + (R_B \times L) = \left(wL \times \frac{L}{2}\right)$$

$$R_B = \frac{3wL}{8} \text{ (check)}$$

Note that the reaction at the fixed support can now be obtained by considering vertical equilibrium in the normal way.

It is worth considering what would happen to the above propped cantilever if either of the two end supports were subjected to settlement. If the propped end settled, the value of the reaction R_B would reduce, with the reaction at A increasing by the same amount (in order to continue to satisfy vertical equilibrium). Also the bending moment at the fixed end A would increase. If the settlement continued until the reaction R_B was reduced to zero (i.e. no contact) we would be left with a simple statically determinate cantilever. If the fixed support settled relative to the propped end then opposite changes would occur. This illustrates a major difference between determinate and indeterminate structures, because, of course, with determinate structures the values of support reactions and bending moments do not change with support settlements.

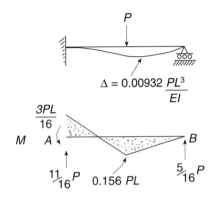

Figure 15.2b
Propped cantilever with central point load

Example 15.1

A propped cantilever has a span of 5.5 metres and supports a design load in the form of a point load at mid-span of 75 kN. Determine the dimensions of a suitable mild steel standard universal beam if the beam is assumed to have full lateral restraint.

Solution

This is a standard case as shown in *figure 15.2b*. As a beam of uniform cross-section will be used throughout, we are only interested in the maximum bending moment which occurs at the fixed support:

$$M_{max} = \frac{3PL}{16} = \frac{3 \times 750 \times 5.5}{16}$$

$$= 773.4 \text{ kNm}$$

From *table 3.4* a typical yield stress for mild steel is 275 N/mm^2.

From *equation [7.3]*

$$\text{Required } S = \frac{M_{ult}}{\sigma_y} = \frac{773.4 \times 10^6}{275 \times 10^3}$$

$$= 2812 \text{ cm}^3$$

From *Appendix*
Try 533 × 210 × 122 kg/m universal beam
(S = 3200 cm³, D = 544.6 mm, t = 12.8 mm)

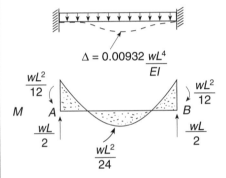

Figure 15.2c
Fully-fixed beam with uniformly distributed load

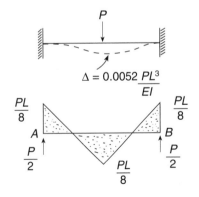

Figure 15.2d
Fully-fixed beam with central point load

Check shear

The maximum shear force will be equal to the maximum reaction force:

from *figure 15.2b* $R_A = \dfrac{11}{16} P = 0.6875 \times 750 = 516 \text{ kN}$

$$\text{Average shear stress} = \frac{516 \times 10^3}{544.6 \times 12.8} = 74.0 \text{ N/mm}^2$$

$$(< 165 \text{ N/mm}^2)$$

Answer   Use 533 × 210 × 122 kg/m universal beam

15.5 Upper and lower bound theorems

To provide additional insight into the nature of indeterminate structures it is useful to consider the upper and lower bound theorems. The terms **upper bound** and **lower bound** actually refer to the upper and lower limits on the load that will cause a structure to collapse. Even if we cannot evaluate an exact collapse load for a structure it is often useful to know its upper and lower bounds. These methods are only appropriate for structural members and materials that fail in a fully plastic manner, i.e. they are not appropriate for brittle materials or members that would buckle before they reached the full plastic moment. In the case of beams in bending, once a full plastic hinge has formed, it must be capable of sustaining the full plastic moment throughout a large angle of rotation until a complete collapse mechanism develops. The theorems are very general, but expressed here in a form suitable for structures.

The **lower bound theorem** states: *If you can find a stress distribution throughout a structure which is in equilibrium with the external loads and which does not exceed the yield stress, then the loads can be safely carried.*

It may be that the above sounds obvious but it has been subjected to rigorous mathematical proof. In most cases the word 'stress' in the above definition can be replaced by the word 'moment'. When applying the lower bound method the solution is always safe. The structure may carry more load but it cannot fail with less.

Example 15.1 is a lower bound solution, but there are an infinite number of other lower bound solutions that meet the criteria of the theorem. It is not necessary to know the distribution of bending moments given in *figure 15.2b*. We could, for example, propose that the propped cantilever will behave as a simply supported beam (ignoring the fixity), a simple cantilever (ignoring the prop) or any intermediate state – say with an arbitrary hogging fixing moment of 0.2*PL*. These three cases are shown in *figure 15.3*. It can be seen that they all produce different bending moments, and hence would result in different beam sizes if the design procedure of *example 15.1* is followed. What the lower bound theorem tells us, however, is that they are all safe. It therefore makes sense to find the lower bound equilibrium solution that produces the smallest maximum bending moment for design purposes. In the case shown in *figure 15.3c* it is necessary to check to see if the maximum bending moment occurs at the fixed support or in sagging under the load.

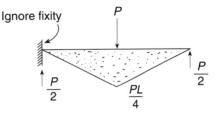

Figure 15.3a
Propped cantilever ignoring fixity

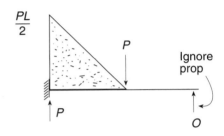

Figure 15.3b
Propped cantilever ignoring prop

(ΣM about fixed support)

anticlockwise moments = clockwise moments

$$0.2PL + (R_\mathrm{B} \times L) = (P \times \frac{L}{2})$$

$$R_\mathrm{B} = 0.3P$$

(ΣM about mid-span) – *see figure 15.3d*

$$M = (0.3P \times \frac{L}{2})$$

$$\boldsymbol{M = 0.15PL}$$

Hence it can be seen that it is 0.2*PL* at the support that should be used in the design of the beam. This value is still, however, slightly greater than the value of bending moment used in *example 15.1* (= 0.1875*PL*) based on the standard case. So our use of the lower bound theorem, although safe, is slightly less economical. However, it is easy to show that if we proposed a bending moment regime where the support moment was equal to the mid-span moment (*figure 15.4*), the value would fall to 0.166*PL* whilst at the same time satisfying the equilibrium requirement. *As an exercise, show that figure 15.4 satisfies equilibrium.* This is now a more economical solution than the standard case.

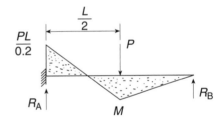

Figure 15.3c
Propped cantilever with assumed fixing moment

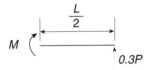

Figure 15.3d
Application of moment equilibrium

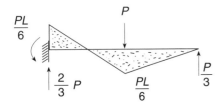

Figure 15.4
The most economic lower bound solution

But how can this be true? How can we justify reducing a support bending moment by over 10%? The answer is that because the beam can plastically deform at the support, but still maintain its load carrying capacity through plasticity, the responsibility for satisfying equilibrium is **redistributed** to other parts of the structure. Load is effectively shed from one part of the structure and taken up by another part. Only plastically deforming redundant structures can do this. This approach is acceptable for evaluating the ultimate collapse loads of structures, although safety factors applied to the loads should ensure that, in reality, the structure stays within its elastic limits and hence does not suffer permanent plastic deformation. Because of the ability of a plastic hinge to continue to support load as it rotates, the ultimate plastic analysis of a structure is immune from the effects of its supports settling (unlike an elastic analysis). One problem that should be borne in mind, however, is that excessive support settlements can cause a complete reversal of moments and hence stresses. Consider a two-span continuous beam for example. Initially the moment over the central support will be hogging, however, after considerable settlement of the central support this may turn into sagging. This could be a problem for a reinforced concrete beam where the main tensile reinforcement was placed in the top face to resist the hogging.

The **upper bound theorem** states: *If a plastic collapse mechanism can be found such that a structure will fail under a given load, then that load estimate is either correct or an over-estimate.*

Unlike the lower bound theorem, the upper bound theorem does not err on the safe side. It involves postulating a failure mechanism for a structure and then determining the load that will force it to collapse. The problem is, however, that we may not have postulated the most critical failure mechanism. There may be another mechanism that would fail at a lower load (but we just have not thought of it!). There may be points between the postulated plastic hinges where the bending moment is higher and hence the yield criterion would be violated.

It is interesting to again examine *example 15.1* and this time subject it to an upper bound approach. Failure mechanisms are generated by introducing plastic hinges into a structure. A number equal to one more than the number of redundancies is required to produce a mechanism. For such a simple structure as this, the critical mechanism is obvious (*figure 15.5*). As a structure collapses work is done by the applied loads (work = force × distance), but this is resisted by work done in the plastic hinges (work = plastic moment × angle of rotation). The required values of the fully plastic moments M_p are thus evaluated by equating the internal and external work done.

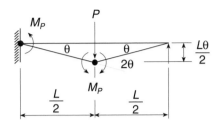

Figure 15.5
Upper bound plastic collapse mechanism

From *figure 15.5* we can see that the plastic hinge at the support is assumed to have rotated through an angle of θ radians. From geometry this implies that the applied load, P, has moved through a distance of $L\theta/2$, and that the central plastic hinge has moved through an angle of 2θ radians. We can thus equate internal and external work done:

$$\text{internal work} = \text{external work}$$

$$(M_\mathrm{p} \times \theta) + (M_\mathrm{p} \times 2\theta) = P \times \frac{L\theta}{2}$$

Hence $\qquad\qquad M_\mathrm{p} = \frac{PL}{6}$

It can be seen that the above upper bound solution equals the lower bound solution shown in *figure 15.4*. This means that we have effectively discovered the highest lower bound and the lowest upper bound, i.e. the true collapse load. With more complex structures and loading cases this is not always possible. Even with the simple propped cantilever problem it would not be obvious where to put the central plastic hinge if the structure was subjected to multiple or UDL loads. It is, however, good practice to draw the bending moment diagram after defining the plastic hinge positions. If, at any point, the diagram indicates that the plastic moment is exceeded, it means that the lower bound theorem has been violated and that the solution is unsafe.

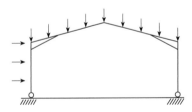

Figure 15.6a
Portal frame with wind loading

Despite the fact that the upper bound approach is intrinsically 'unsafe', it is widely used as economical design method for certain classes of structure where the failure mechanisms are well understood. An example of this is the portal framed factory building, and a typical collapse mechanism for the side-sway load case under wind loads is shown in *figure 15.6*.

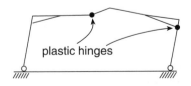

Figure 15.6b
Collapse mechanism

15.6 The computer stiffness method

Before describing the detail of how computers analyse structures it is worth defining the program requirements:

1. The program must be **general purpose**, i.e. it must be flexible enough to analyse a wide range of structures without the need to modify the program itself. Only the data should change.
2. It should provide the designer with information on deflections as well as bending moments, shear forces and stresses.
3. It should be easy to input the data and comprehend the results. In addition it should be easy to modify structural members and loads and re-run the analysis.
4. It should incorporate 'reasonableness' checks on input data to help guard against the 'rubbish in – rubbish out' phenomenon.
5. It should utilise efficient mathematical techniques so that the solution time for large problems is minimised.

The approach, which is most commonly adopted for such programs, is known as the **stiffness method**. Complex structures are built-up from individual elements member-by-member. Firstly it is necessary to consider an individual member.

15.6.1 *The element stiffness matrix*

Figure 15.7a shows a single member, usually referred to as an **element**, which can be assumed to be part of a larger structure. The ends of the element, where it connects to other elements, are called **nodes**. The element is shown within a local axes frame x_L–y_L, where the element is parallel to the x axis. For a 2-D plane-frame problem, each node has three possible movements. It can have displacement components in the x and y directions as well as rotate through an angle, θ. Each of these movements is known as a **degree of freedom**. For end node 1 the degrees of freedom are defined as:

$$\delta_{1x}, \delta_{1y} \text{ and } \theta_1$$

And for end node 2:

$$\delta_{2x}, \delta_{2y} \text{ and } \theta_2$$

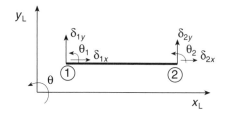

Figure 15.7a
A single element with three degrees of freedom at each end

Now consider that all six degrees of freedom are prevented from moving, i.e. they are rigidly clamped. We will now release only one degree of freedom and see how the application of a load deforms the element. The element has length, L, and cross-sectional area, A. If the clamp restraining δ_{2x} is released (the suffix indicates the node number followed by the direction) and an axial load of P_{2x} is applied (*figure 15.7b*), we know from the simple consideration of stress and strain in *chapter 2* that:

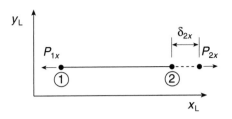

Figure 15.7b
Load applied and deflection permitted in x direction only

$$\text{stress} = \frac{P_{2x}}{A}$$

$$\text{strain} = \frac{\text{stress}}{E} = \frac{P_{2x}}{EA}$$

$$\text{Change in length, } \delta_{2x} = \text{strain} \times L = \frac{P_{2x}L}{EA}$$

Or alternatively
$$P_{2x} = \frac{EA}{L} \times \delta_{2x}$$

The EA/L term is referred to as the **direct stiffness**, k_{22} of the element (in this case in an axial direction). Another effect of applying the load, P_{2x}, is that it induces an equal and opposite force in the clamp that restrains the axial movement at the other end of the element. This is clearly required for horizontal equilibrium of the element.

Hence
$$P_{1x} = -\frac{EA}{L} \times \delta_{2x}$$

$-EA/L$ is referred to as a **cross-stiffness** term, k_{12}. It indicates the effect on node 1 from displacement at node 2.

Now let us consider the effect of releasing one of the other degrees of freedom. Assume δ_{2x} is re-clamped and the displacement δ_{2y} is permitted. A force of P_{2y} is applied (*figure 15.7c*). The equivalent relationship between force and displacement is now given by the following standard case:

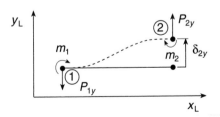

Figure 15.7c
Load applied and deflection permitted in y direction only

$$P_{2y} = \frac{12EI}{L^3} \times \delta_{2y}$$

Where: E = modulus of elasticity

I = second moment of area

P_{2y} causes a force at the other end which is associated with a cross-stiffness term of $-12EI/L^3$.

Hence $$P_{1y} = -\frac{12EI}{L^3} \times \delta_{2y}$$

Examination of *figure 15.7c* indicates that the application of P_{2y} also induces bending moments at the element ends. By considering the equilibrium of the element it is easy to show that these moments are equal to $-6EI/L^2$. These constitute further cross-stiffness terms that affect the rotational degree of freedom at the nodes.

Hence $$m_1 = -\frac{6EI}{L^2} \times \delta_{2y}$$

and $$m_2 = -\frac{6EI}{L^2} \times \delta_{2y}$$

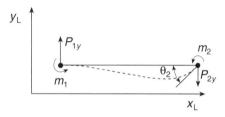

Figure 15.7d

Moment applied and rotation permitted in θ direction only

Lastly, we need to release the rotational degree of freedom at node 2 and subject the element to a moment of m_2 (*figure 15.7d*). Again a standard case gives:

$$m_2 = \frac{4EI}{L} \times \theta_2$$

with cross-stiffness terms:

$$m_1 = \frac{2EI}{L} \times \theta_2$$

$$P_{1y} = \frac{6EI}{L^2} \times \theta_2$$

$$P_{2y} = -\frac{6EI}{L^2} \times \theta_2$$

Now suppose we released all the restraints to the degrees of freedom simultaneously. The various effects from both direct and cross-stiffness terms are simply added together. For node 1 this gives:

$$P_{1x} = \frac{EA}{L}\delta_{1x} - \frac{EA}{L}\delta_{2x}$$

$$P_{1y} = \frac{12EI}{L^3}\delta_{1y} - \frac{12EI}{L^3}\delta_{2y} + \frac{6EI}{L^2}\theta_1 + \frac{6EI}{L^2}\theta_2$$

$$m_1 = \frac{4EI}{L}\theta_1 + \frac{2EI}{L}\theta_2 + \frac{6EI}{L^2}\delta_{2y} - \frac{6EI}{L^2}\delta_{2y}$$

A similar set of equations could be stated for node 2. In effect this would produce a set of six simultaneous equations, where, given the six end forces and moments (P_{1x}, P_{1y}, m_1, etc.) we could solve for the six unknown displacements. Solving large sets of simultaneous equations is tedious and time consuming for humans, however, this is a task for which computers are ideally suited. It is, however, much more convenient to express the equations in matrix form:

$$\begin{Bmatrix} P_{1x} \\ P_{1y} \\ m_1 \\ P_{2x} \\ P_{2y} \\ m_2 \end{Bmatrix} = \begin{bmatrix} \dfrac{EA}{L} & 0 & 0 & -\dfrac{EA}{L} & 0 & 0 \\ 0 & \dfrac{12EI}{L^3} & \dfrac{6EI}{L^2} & 0 & -\dfrac{12EI}{L^3} & \dfrac{6EI}{L^2} \\ 0 & \dfrac{6EI}{L^2} & \dfrac{4EI}{L} & 0 & -\dfrac{6EI}{L^2} & \dfrac{2EI}{L} \\ \dfrac{EA}{L} & 0 & 0 & -\dfrac{EA}{L} & 0 & 0 \\ 0 & -\dfrac{12EI}{L^3} & -\dfrac{6EI}{L^2} & 0 & \dfrac{12EI}{L^3} & -\dfrac{6EI}{L^2} \\ 0 & \dfrac{6EI}{L^2} & \dfrac{2EI}{L} & 0 & -\dfrac{6EI}{L^2} & \dfrac{4EI}{L} \end{bmatrix} \begin{Bmatrix} \delta_{1x} \\ \delta_{1y} \\ \theta_1 \\ \delta_{2x} \\ \delta_{2y} \\ \theta_2 \end{Bmatrix}$$

or more briefly this can be expressed as:

$$\begin{Bmatrix} P_1 \\ P_2 \end{Bmatrix} = \begin{bmatrix} k_{11} & k_{12} \\ k_{21} & k_{22} \end{bmatrix} \begin{Bmatrix} d_1 \\ d_2 \end{Bmatrix}$$

or even

$$\boldsymbol{P} = \boldsymbol{k}\boldsymbol{d}$$

The large matrix above is known as the **member** or **element stiffness matrix**. The following points should be noted:

- It is the basic building block for the stiffness method. The computer evaluates such a stiffness matrix for every individual member in a structure.

- The form of the stiffness matrix is universal for all members. Only the particular values of E, A, I and L will change to suit the properties of individual members.

- The stiffness matrix is symmetrical about the leading diagonal, so the computer only needs to store half of it.

15.6.2 *Assembling the structure stiffness matrix*

We now need to consider how the stiffness matrices for individual members are combined to form an overall stiffness matrix for the whole structure. *Figure 15.8a* shows a portion of a larger structure with four nodes arbitrarily numbered 6, 7, 8 and 9. Note that the axes are now designated x_G and y_G indicating that they now refer to the **global** axes of the whole structure. Now, following a similar process to the individual members, consider all the degrees of freedom to be clamped apart from, say, the x_G direction at node 9 (*figure 15.8b*). It is clear that the force/displacement relationship for node 9 must take account of the stiffness of all the members that meet at 9.

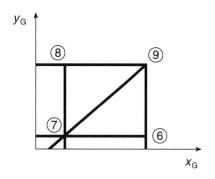

Figure 15.8a

A corner of a large structure in global axes

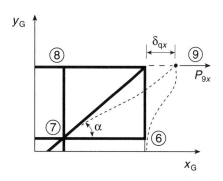

Figure 15.8b

x deflection at node 9 involves deformation of three elements

For member 8–9 the local axes x_L and y_L coincide with the global axes x_G and y_G, so the direct stiffness contribution of that member is simply EA/L_{8-9} as indicated by the member stiffness matrix. However, for the other members meeting at node 9 the local axes do not coincide with the global axes. Member 6–9 is rotated through 90° so that a global x deformation of the structure is equivalent to a local y deformation of that particular member. Hence its stiffness contribution is given by $12EI/L^3_{6-9}$. Member 7–9 lies at an inclination of $\alpha°$ to the global axes and its stiffness will lie somewhere between that of the other two members. In order to evaluate its contribution we need to perform a **transformation**. This can be done by multiplying the normal member stiffness matrix by a **transformation matrix** as follows:

$$\text{Transformation matrix} = \begin{bmatrix} \cos\alpha & \sin\alpha & 0 \\ -\sin\alpha & \cos\alpha & 0 \\ 0 & 0 & 1 \end{bmatrix} = T$$

For a member at an angle of $\alpha°$ to the global axes:

$$\text{Global stiffness} = \begin{bmatrix} T^T & 0 \\ 0 & T^T \end{bmatrix} \begin{bmatrix} k_{11} & k_{12} \\ k_{21} & k_{22} \end{bmatrix} \begin{bmatrix} T & 0 \\ 0 & T \end{bmatrix}$$

Where T^T is the transpose of the above transformation matrix. The above procedure automatically transforms the stiffness matrix of a member from its local axes to the global axes of the whole structure. The combined stiffness equation for each node therefore becomes the sum of the contributions from all the members meeting at that point. The computer assembles the complete stiffness matrix for the whole structure which for a 2-D structure with N nodes has dimensions of $3N \times 3N$. Although this matrix can be very large, if properly constructed it is banded and populated only around the leading diagonal. This feature, combined with the fact that it is symmetrical about the diagonal, means that special numerical techniques can be employed for efficient storage and solution.

3-D structures such as space frames are treated in a similar way to 2-D structures, however, each node now has six degrees of freedom so the structure stiffness matrix has dimensions of $6N \times 6N$.

15.6.3 A typical stiffness program

A modern stiffness computer package contains the following elements:

1. *Pre-processor*
This part of the program assists the user to input the information. In order to perform the analysis the computer needs to know:

- The global co-ordinates of every node
- The end nodes of every member
- The geometric and physical properties of all the members
- The restraints (supports) acting on the structure
- The loads acting on the structure

Much of the above information can be obtained directly from a CAD drawing and most packages are supported by libraries of material properties and standard section properties.

2. *Processor*

This is the part of the program that performs the actual calculation. The basic steps are:

- From the node co-ordinates and member list, determine the length and properties of each member.
- Evaluate the stiffness matrix for each member in its local axes system.
- Transform the stiffness of each member into the global structure axes.
- Assemble the structure stiffness matrix by combining the global stiffnesses of all the members that meet at each node.
- Apply support restraints to the structure (by removing the degrees of freedom that correspond to node fixities).
- Determine the forces and moments at each node from the applied loads.
- From the global stiffness equation, $P_G = k_G d_G$, solve for all the unknown global displacements, d_G.
- For each member, transform the global end displacements into local displacements, d_L, using the transformation matrix, T.
- From the local stiffness equation, $P_L = k_L d_L$, solve for the local end forces and moments, P_L.

3. *Post-processor*

This section takes the results and presents them to the user in an accessible form as follows:

- For each member, combine the member end forces and moments with any loads on the member itself to arrive at the distribution of bending moments, shear forces and displacements throughout the member.
- Evaluate particular stresses or deformations as requested by the user and display them, usually in graphical form.
- Tabulate results as requested and incorporate them into output reports.
- Format the data in a form suitable for transmission to other packages for detailing, costing and production.

15.7 An introduction to the finite element method

The **finite element method** extends the stiffness method so that it can cope with structures that are made from plates and shells as distinct from discrete elements. The method can even analyse complex 3-D solid shapes such as curved arch dams. The finite element method still utilises the basic stiffness equation $P = kd$, however, the problem is that, in dealing with a continuum, the position of the nodal points and the stiffness of the elements between the nodes is no longer obvious as it was with line elements.

Consider the plate shown in *figure 15.9a* that is subjected to in-plane forces. This can be divided into a set of triangular elements joined only at the nodes as shown in *figure 15.9b*. (The method gets its name from the fact that a **continuous** structure is approximated to a **finite** number of discrete elements.) The process of defining the elements is known as **meshing**. Now the aim is to determine the stiffness of each element together with the distribution of strains (and hence stresses) over each element.

15.7.1 *Conforming and non-conforming elements*

You might think that we could obtain the stiffness of the elements from physical tests in the laboratory, i.e. we could make a steel triangle and clamp two of the element corners in a vice and then measure the force required to induce a particular displacement at the remaining node. In fact this would not produce a very effective result. The structural model would be too flexible (resulting in unrealistically high predictions of deformations, strains and stresses). This is because too much relative movement would take place between adjacent edges of neighbouring elements. *Figure 15.10a* shows how, although two adjacent elements are linked at the nodes, the edges have deformed in different ways. This is known as a **non-conforming element**. *Figure 15.10b* shows a **conforming element** in which the deformations of the edges are forced to be compatible. Conformance is usually achieved by forcing the distribution of strains over the element to obey mathematical functions. Two techniques for doing this are known as **displacement functions** and **shape functions**. It follows that the shape of the boundary between two conforming elements should be completely defined by the nodal displacements that lie on that boundary. The displacement of other nodes, not on the boundary, must have no influence.

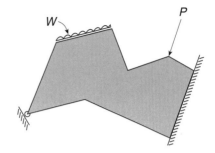

Figure 15.9a
A plate structure with in-plane loads

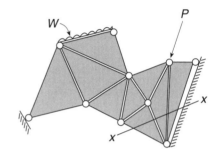

Figure 15.9b
Modelling plate structure with a finite element mesh

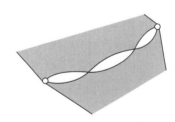

Figure 15.10a
Non-conforming elements

Figure 15.10b
Conforming elements

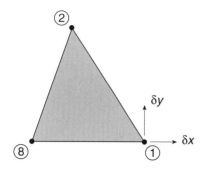

Figure 15.11a

A simple three-node triangular element

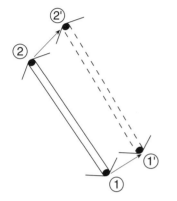

Figure 15.11b

Edges remain straight after deformation

15.7.2 *The three node triangular element*

Many different elements have been formulated with various shapes and complexities. The simplest element for plate problems is the three-node triangle with δ_x and δ_y displacement at each node (*figure 15.11a*). *Figure 15.11b* shows an element boundary after deformation where node 1 has moved to 1′ and node 2 has moved to 2′. The information available to define the shape of the element boundary is clearly limited (simply a change in length), consequently, if it is to be conforming, the element edges must remain straight lines. Similar limitations apply to the complexity of the strain (and hence stress) variations across such a simple element. In fact, for this element, the strain in any direction is constant throughout the element.

With one-dimensional members, such as those considered in the previous stiffness method, the conversion from strain to stress was straightforward:

$$\sigma = E\varepsilon$$

However with two-dimensional structures it is necessary to consider two cases. Thin plate-type structures are subjected to what is known as **plane stress**. In this case the stress through the thickness of the plate, σ_z, is zero. The relationship between stress and strain is:

$$\left\{ \begin{array}{c} \sigma_x \\ \sigma_y \\ \tau_{xy} \end{array} \right\} = \frac{E}{(1 - v^2)} \begin{bmatrix} 1 & v & 0 \\ v & 1 & 0 \\ 0 & 0 & \dfrac{1-v}{2} \end{bmatrix} \left\{ \begin{array}{c} \varepsilon_x \\ \varepsilon_y \\ \gamma_{xy} \end{array} \right\}$$

where v = Poisson's ratio.[1]

The stresses in long prismatic structures, such as straight dams, can also be determined with these elements, and this is known as the **plane strain case**. In this case the longitudinal strain through the structure, ε_z, is zero. The relationship between stress and strain is:

---

[1] When a tensile stress is applied to a structural member it extends owing to longitudinal strain. However, it also gets thinner owing to negative lateral strain. The ratio between longitudinal and lateral strain is known as **Poisson's ratio**. It is a dimensionless elastic constant that for most materials lies between 0 and $\frac{1}{2}$.

$$\left\{\begin{array}{c} \sigma_x \\ \sigma_y \\ \tau_{xy} \end{array}\right\} = \frac{E}{(1+v)(1-2v)} \left[\begin{array}{ccc} 1-v & v & 0 \\ v & 1-v & 0 \\ 0 & 0 & \frac{1-2v}{2} \end{array}\right] \left\{\begin{array}{c} \varepsilon_x \\ \varepsilon_y \\ \gamma_{xy} \end{array}\right\}$$

It is important that it is made clear when inputting the data whether the analysis is based on plane stress or plane strain.

Referring back to *figure 15.9b*, the line marked x–x indicates a cross-section cut through the plate that crosses three elements. If the stresses were plotted along this line, the best solution that this simple finite element can give is the stepped line shown in *figure 15.11c*. Clearly we could get a better approximation to the true stress distribution if we used a greater number of smaller elements (i.e. smaller steps). This means that **in regions where stresses are changing rapidly it is important to use a finer element mesh**. This particularly applies to points of high stress concentration such as the sharp re-entrant corner shown if *figure 15.12*.

15.6.3 *More complex elements*

The next step-up in complexity is to introduce an intermediate side node to the triangular element. This now becomes a six-node triangle (*figure 15.13a*). Displacement of the extra side node now provides enough information to define a quadratic curve for the element boundary (*figure 15.13b*) which is a better approximation to real behaviour.

Higher order elements (i.e. more degrees of freedom) also provide a more accurate variation in strain (and hence stress) throughout each element. Again, referring back to the cross-section shown on *figure 15.9b*, it can be seen in *figure 15.14* that even a simple linear variation in stress across an element can provide a much closer approximation to the true value. Designers are therefore sometimes faced with the choice between using a lot of simple elements or a smaller number of more complex elements. Experience is required to reach the optimum choice for specific problems.

Computer packages can offer a very wide range of different elements to cope with different problems. Some common types are shown in *figure 15.15* and consist of:

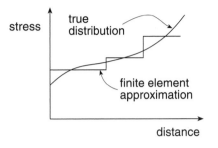

Figure 15.11c
Constant stress across element only approximates to true distribution

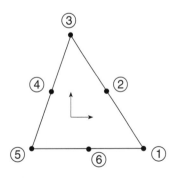

Figure 15.12
Use of finer mesh at points of stress concentration

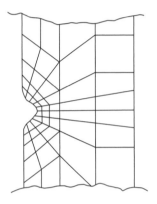

Figure 15.13a
A six-node triangular element

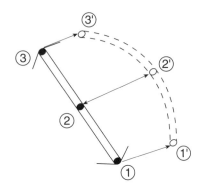

Figure 15.13b

Boundaries curved after deformation

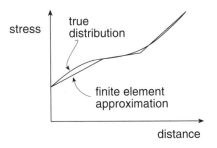

Figure 15.14

Linear variation in stress gives better approximation

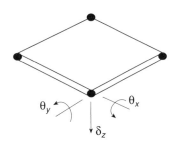

Figure 15.15a

Plate bending element with four nodes

- Plate bending elements to cope with out-of-plane forces and displacements. *Figure 15.15a* shows an element with twelve degrees of freedom. Each corner node contains a δ_z displacement together with two rotations about the X and Y axes. Although this element is non-conforming it has been found to give reasonable results.

- Shell elements for curved 3-D plate structures. With eight nodes and five degrees of freedom per node this element has forty degrees of freedom (*figure 15.15b*).

- Brick elements for modelling solid 3-D structures. This element only has three degrees of freedom per node but with twenty nodes that makes sixty degrees of freedom (*figure 15.15c*). Large complex problems would quickly result in very large matrices.

15.7.4 *The problem of non-linear structures*

In all the computer methods considered so far we have assumed a linear relationship between stress and strain. This means that the effects of different loads, together with the effects of the direct stiffness and the cross-stiffness terms can simply be added together to arrive at a final result. This is satisfactory provided the structural material always remains within the linear elastic range. In the case of mild steel this effectively means that the stresses must remain below the yield stress, σ_y. If the objective is to model a structure beyond the yield stress of the material, say in order to investigate a collapse mechanism, then a non-linear approach must be followed. This implies that the value of the modulus of elasticity, E, that appears in all the stiffness matrices, is no longer a constant.

It is obviously necessary to input the nature of the stress/strain relationship if a non-linear analysis is to be carried out. The simplest form is to assume that the material is elastic/perfectly plastic, in which case only the yield stress, σ_y, is required (*figure 15.16*). If a more complex shape is to be followed then it is necessary to provide either an equation or co-ordinates of the desired curve. The normal approach with a non-linear analysis is that the loads are applied in small increments. Following each increment, an analysis is performed and the maximum strain in each element is checked against the stress/strain curve to see if it has gone beyond the linear range. If it has, then a suitable value of stress is inserted. As the increments are increased it is possible to see an increasing number of yielding elements radiating outwards from points of stress concentration. A non-linear analysis essentially requires the calculations to be carried out many times and so can dramatically increase the load on the computer.

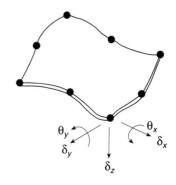

Figure 15.15b
Shell element with curved edges and eight nodes

15.7.5 *Finite element computer packages*

As with stiffness programs, finite element programs require user-friendly pre and post-processors. Commercial finite element (F.E.) packages fall into two main categories:

1. Those that are closely integrated with 3-D drawing and modelling packages. These are intended for preliminary sizing of structural components and are widely used in mechanical engineering design. The F.E. analysis is largely 'black box', with the user generally having no knowledge of the type of element being used. The package usually performs automatic meshing of the elements, although the user often has some control over the coarseness of the mesh. Some indication of the accuracy of the results can be obtained by repeating the analysis with a finer mesh to see if the results are converging.

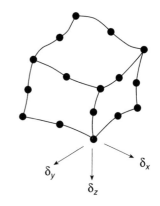

Figure 15.15c
Curved brick element

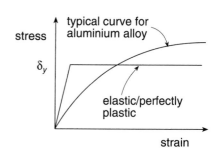

Figure 15.16
Non-linear element properties

2. Specialist F.E. packages provide the user with much more control over element selection and meshing. The facilities for actually modelling the structure are often less helpful than the previous category, although models can generally be imported from CAD packages using particular file formats. Imported models often create problems, however, and require some adjustment. Specialist packages have a significant learning curve but are essential for meaningful analysis of large problems.

As well as dealing with static structural problems, most F.E. computer packages will also perform dynamic vibration and heat flow analyses.

15.8 The dangers of computer analysis

The cheapness and ease of use of modern analysis packages means that they are now accessible to the non-specialist and this carries considerable risk. Several structural failures are now attributable to the use or misuse of computers. An uncritical 'black box' approach can lead to the following problems:

- An inability to appreciate when an answer provided by the computer package is clearly wrong (probably because of errors in input data). An experienced engineer will always subject a solution to reasonableness and equilibrium checks.
- Basing the analysis on a naïve or unrealistic structural model.
- Failing to use sensible loads and support restraints.
- Not appreciating that, for a linear analysis, the stresses must remain below the yield stress.
- Being over-impressed by multi-coloured stress contours as outputs without carefully investigating the data for critical points.
- Attempting to use a single model for both overall structural behaviour and detailed connection design.
- Expecting a complex computer analysis to compensate for fundamental shortfalls in a basic structural concept.
- Failing to take advantage of symmetry to simplify problems.
- Lacking understanding of what is really significant in the output data.
- The fact that the trend is towards more integrated computer packages, where data is passed automatically from a CAD package to an F.E package and then onto a manufacturing package means that the dangers of input errors or poor modelling practice are less likely to be detected.

15.9 Summary of key points from chapter 15

1. **Statically indeterminate** structures contain one or more **redundancies**. These are extra supports or members that make it impossible to analyse the structure using simple equilibrium.
2. Many books contain tables providing values of peak bending moment and deflection for commonly occurring standard cases.

3. The distribution of elastic bending moments and shear forces within a redundant structure change if supports settle.

4. For structures that fail through the formation of fully plastic hinges we can use plasticity theory to predict upper and lower bounds to the collapse load.

5. **Lower bound** solutions satisfy equilibrium without violating the yield criterion and are always **safe.**

6. **Upper bound** solutions propose a failure mechanism and **may not be safe**, but provide an economic means of designing well-understood structures such as redundant portal frames.

7. The **stiffness method** is the main approach used for computer analysis of structures. The basis of the method is understanding the relationship between end forces and displacements for each structural member in the form $P = kd$, where k is the stiffness matrix.

8. The overall structure stiffness matrix is then assembled by summing the stiffness contributions from all the members meeting at each **node**. Solution of a large set of simultaneous equations reveals the unknown nodal displacements which can then be used to evaluate individual member forces.

9. The **finite element method** is related to the stiffness method but can handle continuous structures such as plates, shells and solid objects.

10. The F.E. process involves imposing an artificial **mesh** of elements throughout a structure. There are many elements to choose from and a problem may require many simple elements or fewer complex elements.

11. A finer mesh should be used in areas of high stress concentration

12. Many dangers can arise from inexperienced designers using computer programs as 'black boxes'.

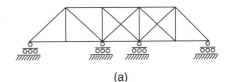

(a)

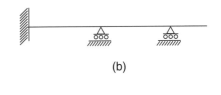

(b)

15.10 Exercises

E15.1   Determine the number of redundancies for the statically indeterminate structures shown in *figure 15.17a–c*
a) 3, b) 2, c) 2.

E15.2   *Figure 15.18* shows a redundant beam structure subjected to a design load of 300 kN. The support bending moments have

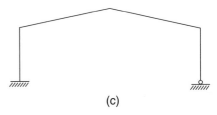

(c)

Figure 15.17
Statically indeterminate structures

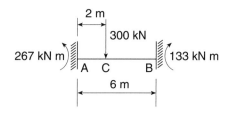

Figure 15.18
Fixed beam with off-centre point load

been inserted from a standard case. By consideration of simple equilibrium, evaluate the support reactions, and peak sagging bending moment. Determine the dimensions of a suitable standard universal beam if the yield stress is 280 N/mm^2.

$R_A = 222$ kN, $R_B = 78$ kN, $M_C = 177$ kNm
Use 406 × 178 × 54 kg/m universal beam

E15.3   For the case shown in *figure 15.18*, carry out an upper bound plastic analysis, and determine the value of the ultimate plastic collapse moment M_p. Establish the dimensions of a suitable standard universal beam (using the same yield stress) and determine the % weight saving compared to E15.2.

$M_p = 200$ kNm
Use 406 × 140 × 39 kg/m universal beam
27.8% saving

E15.4   Outline the basic steps that a stiffness computer program will perform to analyse a large indeterminate structure. List four dangers arising from inexperienced use of computer programs.
See text sections 15.5.3 and 15.7

Appendix

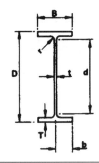

Universal beams
Dimensions and properties of some
standard rolled steel sections

Serial size	Mass per metre	Depth of section D	Width of section B	Thickness		Second moment of area I	Minimum radius of gyration r_{min}	Elastic mod. x–x Z_{x_3}	Plastic mod. x–x S_{x_3}	Torsional index x	Area of section A
				Web t	Flange T						
mm	kg	mm	mm	mm	mm	cm^4	cm	cm^3	cm^3		cm^2
838 × 292	226	850.9	293.8	16.1	26.8	340 000	6.27	7990	160	35.0	289.0
	194	840.7	292.4	14.7	21.7	279 000	6.06	6650	7650	41.6	247.0
	176	834.9	291.6	14.0	18.8	246 000	5.90	5890	6810	46.5	224.0
762 × 267	197	769.6	268.0	15.6	25.4	240 000	5.71	6230	7170	33.2	251.0
	147	753.9	265.3	12.9	17.5	169 000	5.39	4480	5170	45.1	188.0
686 × 254	170	692.9	255.8	14.5	23.7	170 000	5.53	4910	5620	31.8	217.0
	140	683.5	253.7	12.4	19.0	136 000	5.38	3990	4560	38.7	179.0
	125	677.9	253.0	11.7	16.2	118 000	5.24	3480	4000	43.9	160.0
533 × 210	122	544.6	211.9	12.8	21.3	76 200	4.67	2800	3200	27.6	156.0
	101	536.7	210.1	10.9	17.4	61 700	4.56	2300	2620	33.1	129.0
	82	528.3	208.7	9.6	13.2	47 500	4.38	1800	2060	41.6	104.0
457 × 191	98	467.4	192.8	11.4	19.6	45 700	4.33	1960	2230	25.8	125.0
	82	460.2	191.3	9.9	16.0	37 100	4.23	1610	1830	30.9	105.0
	67	453.6	189.9	8.5	12.7	29 400	4.12	1300	1470	37.9	85.4
406 × 178	74	412.8	179.7	9.7	16.0	27 300	4.03	1320	1500	27.6	95.0
	60	406.4	177.8	7.8	12.8	21 500	3.97	1060	1190	33.9	76.0
	54	402.6	177.6	7.6	10.9	18 600	3.85	925	1050	38.5	68.4
406 × 140	46	402.3	142.4	6.9	11.2	15 600	3.02	778	888	38.8	59.0
	39	397.3	141.8	6.3	8.6	12 500	2.89	627	721	47.4	49.4
305 × 127	48	310.4	125.2	8.9	14.0	9500	2.75	612	706	23.3	60.8
	42	306.6	124.3	8.0	12.1	8140	2.70	531	610	26.5	53.2
	37	303.8	123.5	7.2	10.7	7160	2.67	472	540	29.6	47.5
254 × 146	43	259.6	147.3	7.3	12.7	6560	3.51	505	568	21.1	55.1
	37	256.0	146.4	6.4	10.9	5560	3.47	434	485	24.3	47.5
	31	251.5	146.1	6.1	8.6	4440	3.35	353	396	29.4	40.0
254 × 102	28	260.0	102.1	6.4	10.0	4010	2.22	308	353	27.5	36.2
	25	257.0	101	6.1	8.4	3410	2.14	265	306	31.4	32.2
	22	254.0	101.6	5.8	6.8	2870	2.05	226	262	35.9	28.4

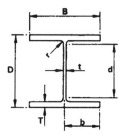

Universal columns

| Serial size | Mass per metre | Depth of section | Width of section | Thickness | | Second moment of area | Minimum radius of gyration | Elastic mod. | Plastic mod. | Torsional index | Area of section |
| | | **D** | **B** | Web **t** | Flange **T** | I_x | r_{min} | Z_y | S_x | x | **A** |
mm	kg	mm	mm	mm	mm	cm^4	cm	cm^3	cm^3		cm^2
305 × 305	283	365.3	321.8	26.9	44.1	78 800	8.25	1530	5100	7.65	360
	137	320.5	308.7	13.8	21.7	32 800	7.82	691	2300	14.1	175
	97	307.8	304.8	9.9	15.4	22 200	7.68	477	1590	19.3	123
254 × 254	167	289.1	264.5	19.2	31.7	29 900	6.79	741	2420	8.49	212
	107	266.7	258.3	13.0	20.5	17 500	6.57	457	1490	12.4	137
	89	260.4	255.9	10.5	17.3	14 300	6.52	379	1230	14.4	114
203 × 203	86	222.3	208.0	13.0	20.5	9460	5.32	299	979	10.2	110
	60	209.6	205.2	9.3	14.2	6060	5.19	199	652	14.1	75.8
	52	206.2	203.9	8.0	12.5	5260	5.16	174	568	15.8	66.4
	46	203.2	203.2	7.3	11.0	4560	5.11	151	497	17.7	58.8
152 × 152	37	161.8	154.4	8.1	11.5	2220	3.87	91.8	310	13.3	47.4
	30	157.5	152.9	6.6	9.4	1740	3.82	73.1	247	16.0	38.2
	23	152.4	152.4	6.1	6.8	1260	3.68	52.9	184	20.4	29.8

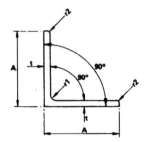

Equal angles

Serial size $A \times A$ mm	Thickness t mm	Mass per metre kg	Minimum radius of gyration r_{min} cm	Area of section A cm^2	Serial size $A \times A$ mm	Thickness t mm	Mass per metre kg	Minimum radius of gyration r_{min} cm	Area of section A cm^2
100 × 100	15	21.8	1.93	27.9	80 × 80	10	11.90	1.55	15.1
	12	17.8	1.94	22.7		8	9.63	1.56	12.3
	8	12.2	1.96	15.5		6	7.34	1.57	9.4
90 × 90	10	13.4	1.75	17.1	60 × 60	10	8.69	1.16	11.1
	8	10.9	1.76	13.9		8	7.09	1.16	9.03
	7	9.61	1.77	12.2		6	5.42	1.17	6.91
	6	8.3	1.78	10.6		5	4.57	1.17	5.82

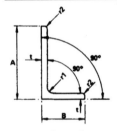

Unequal angles

Serial size $A \times B$ mm	Thickness t mm	Mass per metre kg	Minimum radius of gyration r_{min} cm	Area of section A cm^2	Serial size $A \times B$ mm	Thickness t mm	Mass per metre kg	Minimum radius of gyration r_{min} cm	Area of section A cm^2
150 × 75	15	24.8	1.58	31.6	100 × 65	10	12.3	1.39	15.6
	112	20.2	1.59	25.7		8	9.94	1.40	12.7
	10	17.0	1.60	21.6		7	8.77	1.40	11.2
100 × 75	12	15.4	1.59	19.7	80 × 60	8	8.34	1.27	10.6
	10	13.0	1.59	16.6		7	7.36	1.28	9.38
	8	10.6	1.60	13.5		6	6.37	1.29	8.11

Square hollow sections

Serial size	Thickness	Mass per metre	Area of section	Second moment of area	Radius of gyration	Elastic modulus	Plastic modulus
$D \times D$	t		A	I	r	Z	S
mm	mm	kg	cm^2	cm^4	cm	cm^4	cm^3
60×60	3.0	5.34	6.80	36.6	2.32	12.2	14.5
	4.0	6.97	8.88	46.1	2.28	15.4	18.6
	5.1	8.54	10.9	54.4	2.24	18.1	22.3
	6.2	10.5	13.3	63.4	2.18	21.1	26.6
	8.3	12.8	16.3	72.4	2.11	24.1	31.4
80×80	3.0	7.22	9.20	90.6	3.14	22.7	26.5
	5.0	11.7	14.9	139	3.05	34.7	41.7
	6.1	14.4	18.4	165	3.00	41.3	50.5
	8.2	17.8	22.7	194	2.92	48.6	60.9
100×100	4.0	12.0	15.3	234	3.91	46.8	54.9
	5.0	14.8	18.9	283	3.87	56.6	67.1
	6.3	18.4	23.4	341	3.81	68.2	82.0
	8.0	22.9	29.1	408	3.74	81.5	99.9
	10.0	27.9	35.5	474	3.65	94.9	119

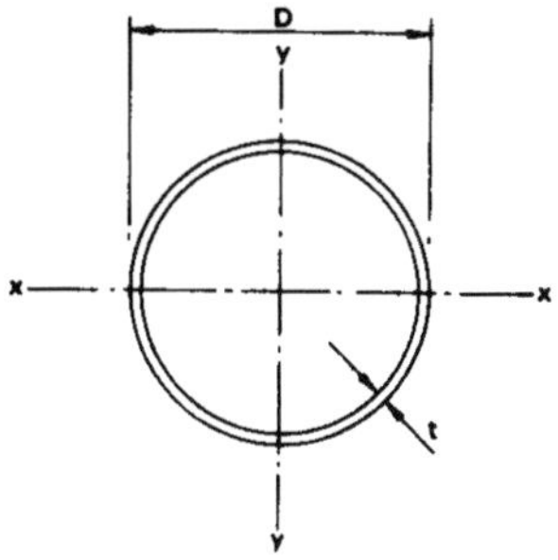

Circular hollow sections

Outside diameter **D** mm	*Thickness* **t** mm	*Mass per metre* kg	*Second moment of area* **I** cm^4	*Radius of gyration* **r** cm	*Elastic Mod.* **Z** cm^3	*Plastic Mod.* **s** cm^3	*Area of section* **A** cm^2
168.3	8.0	31.6	1300	5.67	154.0	206.0	40.3
	5.0	20.1	856	5.78	102.0	133.0	25.7
114.3	6.3	16.8	313	3.82	54.7	73.6	21.4
	5.0	13.5	257	3.87	45.0	59.8	17.2
76.1	5.0	8.77	70.9	2.52	18.6	25.3	11.2
	4.0	7.11	59.1	2.55	15.5	20.8	9.06
60.3	5.0	6.82	33.5	1.96	11.1	15.3	8.69
	4.0	5.55	28.2	2.00	9.34	12.7	7.07
42.4	4.0	3.79	8.99	1.36	4.24	5.92	4.83
	2.6	2.55	6.46	1.41	3.05	4.12	3.25
33.7	4.0	2.93	4.19	1.06	2.49	3.55	3.73
	2.6	1.99	3.09	1.10	1.84	2.52	2.54
21.3	3.2	1.43	0.77	0.65	0.72	1.06	1.82

Further reading

Principal national standards referred to:

BS 5268: Part 2: 2002   Structural use of timber
BS 5628: Part 1: 1992   Structural use of unreinforced masonry
BS 5950: Part 1: 2000   Structural use of steelwork in building
BS 5950: Part 3: 1990   Design in composite construction
BS 6399: Part 1: 1996   Design loading for buildings – dead and imposed loads
BS 6399: Part 2: 1997   Wind loads
BS 6399: Part 3: 1988   Design loading for roofs
BS 8004: 1986   Foundations
BS 8110: Part 1: 1997   Structural use of concrete
BS 8118: 1991   Structural use of aluminium

The following are pre-standards for provisional application
Eurocode 2: BS ENV 1992   Design of concrete structures
Eurocode 3: BS ENV 1993   Design of steel structures
Eurocode 4: EN 1994   Design of composite steel and concrete structures
Eurocode 5: BS ENV 1995   Design of timber structures
Eurocode 6: BS ENV 1996   Design of masonry structures

Books for further reading:

Engineering Materials – An introduction to their properties and applications, M.F. Ashby and D.R.H. Jones, Second Edition, 1996, Butterworth Heinemann

Civil Engineering Materials, Edited by N. Jackson and R.K. Dhir, Fifth Edition, 1996, Palgrave Macmillan

Foundation Design and Construction, M.J. Tomlinson, Seventh Edition, 2001, Prentice Hall

Reinforced Concrete Design, W.H. Mosley and J.H. Bungey, Fifth Edition, 199p, Palgrave Macmillan

Limit States Design of Structural Steelwork, D. Nethercot, Third Edition, 2001, Spon Press

Steelwork Design Guide to BS 5950: Part 1: 2000, Volume 1, Section Properties and Member Capacities, Sixth Edition, 2001, The Steel Construction Institute

Structural Steelwork, W. Mackenzie, 1998, Palgrave Macmillan

Soil Mechanics – Principles and Practice, G.E. Barnes, Second Edition, 2000, Palgrave Macmillan

Basic Solid Mechanics, D.W.A. Rees, 1997, Palgrave MacMillan

Structural Mechanics, J.A. and R. Hulse, Second Edition, 2000, Palgrave Macmillan

Index